Springer Proceedings in Earth and Environmental Sciences

The series Springer Proceedings in Earth and Environmental Sciences publishes proceedings from scholarly meetings and workshops on all topics related to Environmental and Earth Sciences and related sciences. This series constitutes a comprehensive up-to-date source of reference on a field or subfield of relevance in Earth and Environmental Sciences. In addition to an overall evaluation of the interest, scientific quality, and timeliness of each proposal at the hands of the publisher, individual contributions are all refereed to the high quality standards of leading journals in the field. Thus, this series provides the research community with well-edited, authoritative reports on developments in the most exciting areas of environmental sciences, earth sciences and related fields.

Ileana Calotescu • Costin Ioan Coşoiu
Editors

Proceedings of the Third National Conference on Wind Engineering

September 11-13, 2024, Bucharest, Romania

Springer

Editors

Ileana Calotescu
Department Steel Structures, Management and Engineering Graphics
Technical University of Civil Engineering Bucharest
Bucharest, Romania

Costin Ioan Coşoiu
Department of Hydraulics, Sanitary Engineering and Environmental Protection
Technical University of Civil Engineering Bucharest, Aerodynamics and Wind Engineering Laboratory "Constantin Iamandi"
Bucharest, Romania

ISSN 2524-342X ISSN 2524-3438 (electronic)
Springer Proceedings in Earth and Environmental Sciences
ISBN 978-3-032-09670-8 ISBN 978-3-032-09671-5 (eBook)
https://doi.org/10.1007/978-3-032-09671-5

This Springer imprint is published by the registered company Springer Nature Switzerland AG
The registered company address is: Gewerbestrasse 11, 6330 Cham, Switzerland

Preface

The Third National Conference on Wind Engineering (3NCWE) was held on September 11–13, 2024, at the Novotel Bucharest City Centre, Romania. The event was co-organized by the Romanian Association for Wind Engineering (ARIV), the Technical University of Civil Engineering Bucharest (UTCB), and the Faculty of Geography, University of Bucharest (UB). This was the third edition of a conference series that began more than 26 years ago, in 1998, with its inaugural meeting.

The second edition took place in 2019, shortly after the formal establishment of the Romanian Association for Wind Engineering, marking the start of its sustained development, and with the intention of becoming a recurring event. In this context, the conference is of significant importance for the local wind engineering community, acting as a forum for the exchange of ideas, the initiation of collaborations, and the strengthening of connections with the international community.

Reflecting on this mission, the 2024 edition gathered 45 participants, academics, students, and professionals, from 10 countries: Brazil, Canada, China, Germany, Italy, Japan, Romania, Serbia, the United Kingdom, and the United States. The technical program covered a wide range of topics, including aeroelasticity, bluff body aerodynamics, codes, norms, and standards, computational wind engineering, field monitoring, full-scale and wind tunnel measurements, flow–structure interactions, human comfort in the built environment, loads induced by hurricanes, tornadoes, and downbursts, local wind loads on roofing and cladding, stochastic modelling and simulation, wind disaster mitigation, among others. In total, 26 papers were presented, along with four keynote lectures delivered by Professor Bert Blocken, Professor Horia Hangan, Professor Rüdiger Höffer, and Professor Yukio Tamura.

The Organizing Committee honored two of the founding fathers of wind engineering in Romania, Professor Dan Lungu and Professor Constantin Iamandi (1928–2005), by presenting awards in their names to outstanding students attending the conference. Crina Radu from the University of Bucharest received the Dan Lungu Award for Best Paper, while Asad Ullah from Bauhaus-Universität Weimar received the Constantin Iamandi Award for Best Presentation. The corresponding papers are included in this volume.

The editors gratefully acknowledge the contributions of the reviewers and extend their appreciation to all authors for their valuable work. We also thank all participants of the 3NCWE for their role in making the conference a success, as well as the members of the 3NCWE Organizing Committee: Marina Antonescu, Georgeta Bandoc, Adriana Chitez, Andrei-Mugur Georgescu, Mihail Iancovici, Eduard Luca, Florin Pavel, Crina Radu, Anca Tudoricu, and Vlăduț Alexandru Cezar.

Bucharest, Romania

Ileana Calotescu
Costin Ioan Coşoiu

Contents

Estimation of Wind-Induced Fatigue Damage to Beam-Column Welded Joints of Low-Rise Steel Buildings

Yasushi Uematsu, Kosuke Sato, and Taku Matsumoto

Abstract The present paper investigates the wind-induced fatigue damage to the beam-column welded joints of flat-roofed low-rise steel buildings under practical conditions of high winds. First, a wind tunnel experiment is conducted to obtain the time histories of wind pressure coefficients at many points on the building. Then, a fatigue test and a finite element elasto-plastic analysis on the welded joints provide data on the fatigue fracture and the relation between the amplitude of local strain at crack initiation location and the number of loading cycles to fatigue fracture (i.e., fatigue curve). Finally, the wind-induced fatigue damage to the welded joints is evaluated using the fatigue curve and the time histories of wind pressure coefficients obtained from the wind tunnel experiment. The results indicate that the local elasto-plastic behavior at the crack initiation location plays a significant role in evaluating the degree of cumulative fatigue damage to the welded joints.

Keywords Fatigue · Beam-column welded joint · Low-rise steel building · Wind

1 Introduction

In terms of total floor area, about 54% of steel buildings constructed in Japan are classified as low-rise buildings of one or two stories [1]. Because they are lightweight, wind loads dominate their structural design in many cases. The

Y. Uematsu (✉)
Tohoku University, Sendai, Japan
e-mail: yasushi.uematsu.d8@tohoku.ac.jp

K. Sato
Mie University, Tsu, Japan
e-mail: sato@arch.mie-u.ac.jp

T. Matsumoto
Nippon Steel Corporation, Tokyo, Japan
e-mail: matsumoto.ba2@jp.nipponsteel.com

I. Calotescu, C. I. Coşoiu (eds.), *Proceedings of the Third National Conference on Wind Engineering*, Springer Proceedings in Earth and Environmental Sciences,
https://doi.org/10.1007/978-3-032-09671-5_1

dynamic load effect of fluctuating wind pressures on the structure is one of the major concerns for structural engineers when designing these buildings. The duration of high winds during a strong typhoon is so long that fatigue of structural members may occur, which cannot be reproduced by conventional static loading tests. Fatigue analysis using the time histories of wind pressure coefficients on the building, which can be obtained from a wind tunnel experiment, is needed to accurately evaluate the degree of cumulative fatigue damage to the structural members (including the joints).

With an increase in slenderness together with a decrease in structural damping, steel structures have become more susceptible to dynamic wind actions. Turbulent winds cause random vibrations in the along-wind direction to the structures. The resultant stresses induced in the structural members increase as the wind speed increases. The along-wind vibrations are accompanied by mean responses due to mean wind forces, which may play a significant role in fatigue damage to the structural members. Slender structures are subjected to cyclic loads in the across-wind direction due to the generation of Karman vortices. Such dynamic wind loads may cause fatigue damage to the members of steel structures, particularly to the welded joints. Many researchers have investigated the wind induced fatigue of wind sensitive slender structures such as traffic-signal support structures and lighting columns (Repetto and Solari [2, 3]).

The welded joints of steel members, e.g. beam-column joints and column bases, are subjected to high stress. Furthermore, the welded joints have lower fatigue strength than the base metal [4]. Therefore, they are more susceptible to wind-induced fatigue damage than the other parts. Fatigue characteristics of welded joints of steel members were experimentally investigated by Cheng et al. [5], Chen et al. [6], and Karabulut and Rossi [7]. However, most of these studies dealt with high-cycle fatigue with low stresses. By comparison, Narihara et al. [8], Tanuma et al. [9] and Ogawa et al. [10] experimentally investigated the fatigue strength of beam-column welded joints of steel buildings using large-scale or full-scale specimens, in which they considered low-cycle fatigue as well as high-cycle fatigue. However, the targets of their studies were high-rise buildings. To the authors' best knowledge, no study has been done on the fatigue damage to the welded joints of low-rise steel buildings constructed of wide flanges of about 10 mm thickness.

In Japan, low-rise steel buildings are usually designed based on the allowable unit stresses, which are specified in the Building Standard Law of Japan [11]. The damage-limit design wind speed is a 50-year recurrence 10-min mean wind speed at the mean roof height. The short-term allowable unit stress corresponds to the lower limit of yield stress of the material. When the wind speed is close to the design wind speed, the stresses induced in the structural members approach the yield stress. In such a case, not only high-cycle fatigue but also low-cycle fatigue may occur, resulting in the collapse of the structure due to the accumulation of fatigue damage.

The purpose of the present paper is to propose an evaluation method of wind-induced fatigue damage to the beam-column welded joints of low-rise steel buildings considering a wide range of strain from 'elastic' to 'plastic.' First, a wind tunnel experiment is conducted to obtain the time histories of wind pressure

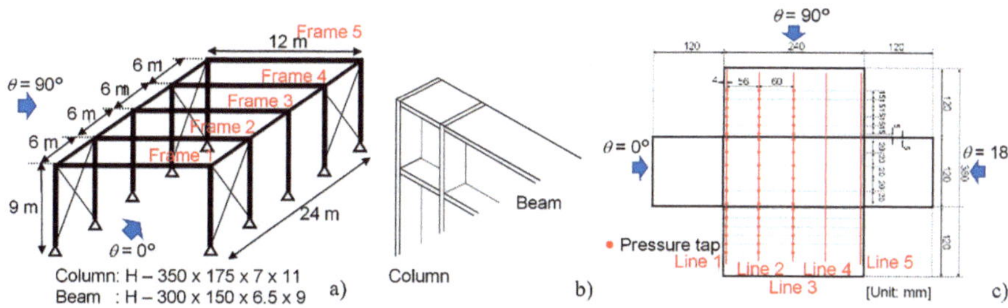

Fig. 1 Investigated building and its wind-tunnel model: (**a**) structural system and dimensions, (**b**) beam-column welded joint and (**c**) pressure tap location

coefficients at many points on the building. Then, a fatigue test is conducted to obtain the fatigue curve for the welded joints in a wide range of strain. A finite element elasto-plastic analysis for the test specimens is also conducted to obtain the strain distribution in the specimen. A model of the relation between the local strain at the crack initiation location and the number of loading cycles to fatigue fracture (fatigue curve) is proposed. Finally, we evaluate the degree of cumulative fatigue damage to the welded joints of low-rise steel buildings using the fatigue curve and the time histories of wind pressure coefficients obtained from the wind tunnel experiment.

2 Investigated Building

Figure 1a shows the investigated building, which consists of moment frames. The material is 'SN400B' [12] or 'SS400' [13], which has the design strength (F) of 235 N/mm^2, corresponding to the lower limit of yield stress. The columns and beams are rigidly welded together (Fig. 1b). The column base is pin-jointed to the foundation. The dimensions of the members are determined based on the usual allowable stress design, specified in the Building Standard Law of Japan [11]. That is, the stresses induced in the members under dead load and wind load shall be less than the specified short-term allowable stresses. The dead load is assumed to be 200 N/m^2. The wind load per unit area is provided by the product of design velocity pressure q_H, wind force coefficient C_f, and gust effect factor G_f. The velocity pressure is given by $0.5\rho V_H^2$, where ρ = air density and V_H = 50-year recurrence 10-min mean wind speed at the roof height H. The '*reference wind speed*' V_0, which determines V_H, is set to 35 m/s, and the terrain category is assumed to be 'III' (suburban). The values of C_f and G_f are provided in the Specification No. 1454 of the Ministry of Construction, Japan [11]. Note that the Building Standard Law of Japan is used for determining the structural members, while the wind loads for estimating the fatigue damage to the beam-column welded joints are evaluated by the AIJ Recommendations for Loads on Buildings (AIJ-RLB) [14].

3 Wind Tunnel Experiment

3.1 Experimental Facilities and Procedure

The wind tunnel model is made at a geometric scale (λ_L) of 1/100. The development diagram of the model is illustrated in Fig. 1c. The model height H can be adjusted by moving the model up and down in a range from 0 to 90 mm, but it is set to 90 mm in the present study. Pressure taps of 0.6 mm internal diameter are installed on Lines 1–3, which correspond to Frames 1–3 (see Fig. 1a). Each line has 18 pressure taps to measure the wind pressure distribution on the line. There are no pressure taps on Lines 4 and 5. However, we can obtain the pressure data on these lines from the data on Lines 2 and 1, respectively, considering the symmetry of the model.

The wind tunnel flow is a turbulent boundary layer generated by using spires and roughness blocks distributed on the wind tunnel floor. Fig. 2 shows the profiles of the mean wind speed U_z, normalized by the value U_{600} at a height of $z = 600$ mm, and the turbulence intensity I_z at the model's center with no model in position. The power-law exponent α for the mean wind speed profile is about 0.21. The intensity (I_H) and integral scale of turbulence (L_{xH}) at the roof height H are about 0.15 and 0.2 m, respectively. The design wind speed U_H at the roof height H is determined based on AIJ-RLB [14]. 'Basic wind speed' U_0 is assumed to be 35 m/s as a representative value for the Main Island of Japan. Assuming that the return period of design wind speed is 50 years and the 'Terrain Category' is III (suburban), U_H is calculated as 26.8 m/s. In the wind tunnel experiment, the mean wind speed at the roof height H is set to 8 m/s. Thus, the velocity scale (λ_V) of the wind tunnel flow is calculated as 1/3.35, resulting in the time scale (λ_T) of 1/29.9. In Sect. 7, where we investigate the fatigue characteristics at higher wind speeds ($U_0 = 40$ and 45 m/s), the velocity and time scales are calculated in the same way. The Reynolds number Re, defined by $Re = U_H H/\nu$, with ν being the kinematic viscosity of air, is about 4.8×10^4. The blockage ratio Br, defined as the ratio of the model's vertical cross-sectional area to that of the wind tunnel, is about 1.72% at most. The values of Re and Br satisfy the requirements of the ASCE Wind Tunnel Testing for Buildings and Other Structures [15], i.e., $Br < 5\%$ and $Re > 1.1 \times 10^4$.

The wind direction θ is varied from 0° to 180° at an increment of 15° (see Fig. 1). The pressures on Lines 2 and 1 for $\theta = 90° - 180°$ are used for those on Lines 4 and 5 for $\theta = 90° - 0°$. The wind pressures at all pressure taps are sampled simultaneously at a sampling rate of 500 Hz for a sampling period of 20.1 s which corresponds to 600 s at full scale. A low-pass filter with a cut-off frequency of 300 Hz is used to remove high-frequency electrical noise. The distortion of the measured fluctuating wind pressures due to the tubing is compensated by using the frequency response function of the measuring system in the frequency domain. The measurement is repeated 10 times under each experimental condition. The wind pressure $P_{e,j}(t)$ at pressure tap j and time t is normalized to a wind pressure coefficient $C_{pe,j}(t)$ as follows:

Fig. 2 Profiles of mean wind speed U_z and turbulence intensity I_z of the wind tunnel flow

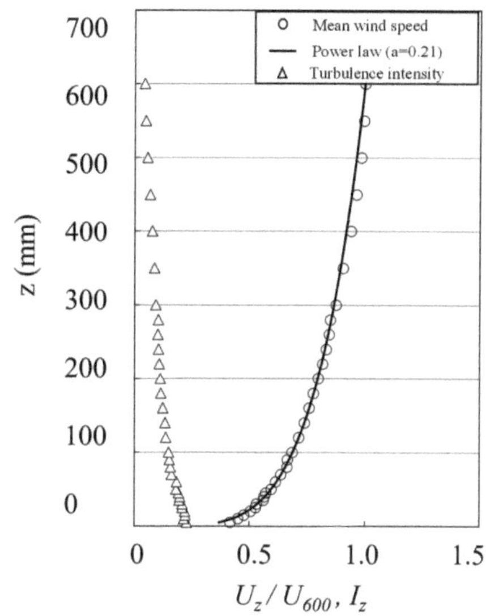

$$C_{pe,j}(t) = \left(P_{e,j}(t) - P_s\right)/q_H \tag{1}$$

where P_s = static pressure; and q_H = mean velocity pressure at the roof height H.

3.2 Experimental Results

The distributions of mean wind pressure coefficients $\overline{C}_{pe}$ on the building at $\theta = 0°$, 45° and 90° are illustrated in Fig. 3. When $\theta = 0°$, large suctions act on the load-bearing area of Frame 1. The roof's dead loads offset these suctions. The distributions of mean wind forces and dead loads are both symmetric with respect to the centerline of the frame. Therefore, the bending moment induced in this frame is not expected to be so large. When $\theta = 90°$, on the other hand, the windward and leeward walls are subjected to positive and negative wind pressures, respectively. Large suctions act on the roof near the windward edge. As a result, the mean wind pressure distribution on the beam shows a marked asymmetry with respect to the centerline of the frame. Thus, it is thought that the bending moments induced in the frame become large. In diagonal winds (e.g., $\theta = 45°$), very large suctions act on the roof near the windward edges due to the generation of conical vortices. In particular, large suctions act on the load-bearing area of Frame 1. However, the load-bearing width of this frame is one-half that of the inner frames. Although the suctions acting on the load-bearing areas of Frames 2 and 3 are also high, the area of such high suctions is limited. These results imply that the most critical situation may occur on the inner frames when $\theta = 90°$, in terms of the bending moments induced in the frame.

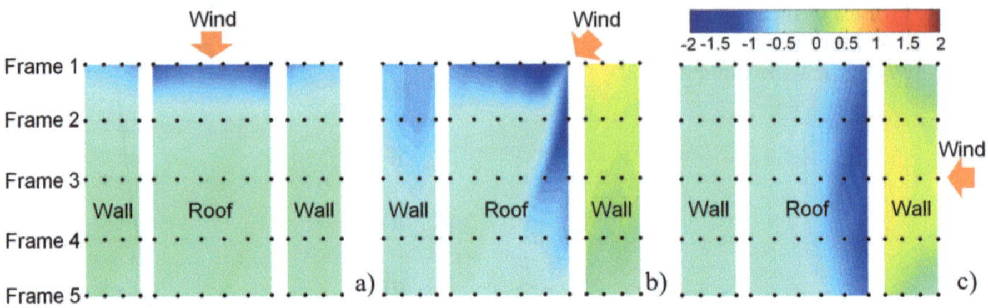

Fig. 3 Distributions of mean wind pressure coefficients on the side walls and roof: (**a**) $\theta = 0°$, (**b**) $\theta = 45°$ and (**c**) $\theta = 90°$

4 Possibility of Wind-Induced Fatigue Damage to the Frame

The wind force per unit area (net wind pressure) on the building is provided by the difference between external and internal pressures. In this paper, the internal pressure is assumed to be zero for simplicity. In this case, the wind force coefficient $C_{f,j}(t)$ coincides with the external wind pressure coefficient $C_{pe,j}(t)$. Furthermore, it is assumed that the wind-induced response is regarded as 'quasi-static.' That is, the effect of resonance on the dynamic response of the frame can be neglected. Consequently, the wind-induced load effect $R(t)$ at time t is provided by the following equation:

$$R(t) = q_H \sum_{j=1}^{n} \alpha_j C_{f,j}(t) A_i \tag{2}$$

where n = number of pressure taps (= 18 in the present case); α_j = load effect influence coefficient of pressure tap j, which is equal to the value of load effect when a unit wind force coefficient is applied to pressure tap j; and A_j = tributary area of pressure tap j.

Figure 4 shows the finite element model (2D model) employed in the present study, which consists of 22 nodes and 21 beam elements. The nodes correspond to the pressure taps on the wind tunnel model, except for the column bases and the beam-column joints. In this study, focus is on the bending moments induced in the frames. Using the time histories of wind pressure coefficients $C_{pe,j}(t)$ and the dead loads, we compute the bending moment $M_k(t)$ at node k (k = 1, ..., 22). Then, we compute the extreme fiber stress $\sigma_{b,k}(t)$ at this node.

A series of dynamic response analyses for all wind directions and all frames indicated that the maximum peak value of $\sigma_{b,k}(t)$ occurred at the windward beam-column joint (Point B) of Frame 2 when $\theta = 90°$. As mentioned in the previous section, large suctions are induced on the roof in diagonal winds (e.g., $\theta = 45°$). However, because the area of such high suctions is limited, the stresses induced in the frames are not so large. Fig. 5 shows an example of the time history of

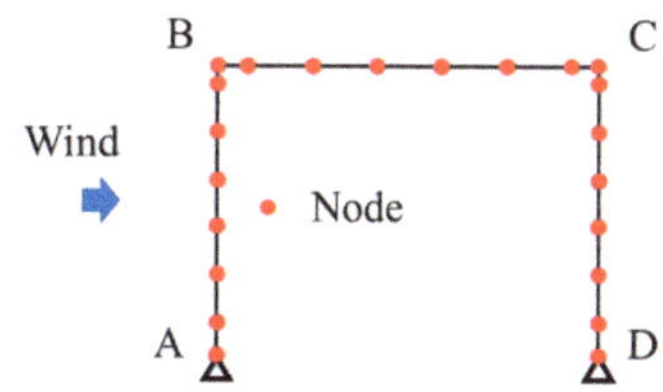

Fig. 4 Finite element model (2D model)

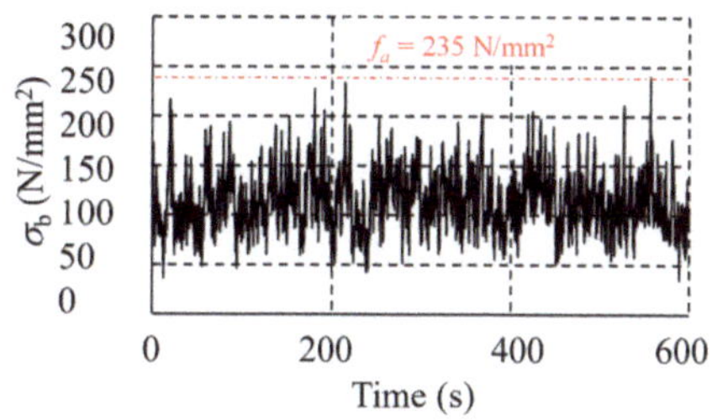

Fig. 5 Time history of the extreme fiber stress σ_b at Point B of Frame 2 when $\theta = 90°$

the extreme fiber stress $\sigma_b(t)$ at Point B of Frame 2. The maximum peak value of $\sigma_b(t)$ almost reaches the short-term allowable stress, i.e., $f_a = 235$ N/mm^2 (= F). Considering that the F value corresponds to the lower limit of yield stress of the material, this feature implies that the maximum stress at the beam-column joint will be close to the yield stress even at the damage-limit design wind speed. According to the Building Standard Law of Japan [11], the safety-limit design wind load is 1.6 times the damage-limit design wind load; the corresponding '*Basic wind speed*' is about 45 m/s. At such a high wind speed, the strain at the welded joint may exceed the yield strain considering that the yield strain of the weld metal is lower than that of the base metal. Therefore, not only high-cycle fatigue but also low-cycle fatigue of the welded joints should be considered appropriately in the wind-resistant structural design of low-rise steel buildings.

5 Fatigue Test on the Beam-Column Welded Joints

5.1 Test Specimen

Gas shielded arc welding is used for welding. The specimens are fabricated from 300 mm wide cruciform welded joints. The thickness of the specimen is 9 mm or 12 mm, and the material is 'SN400B' [11] or 'SS400' [12]. Table 1 summarizes the mechanical properties of the base metal, which are obtained from tensile testing. In the test, we used two or three specimens for each of the four variants. The values presented in the table are the means of the two or three results. The deviation of each value from the mean was less than 1% in most cases and less than 3% at the maximum. Figure 6a shows the dimensions and the mounting position of the instruments for the 9 mm thick specimen. Figure 6b shows the distributions of the

Table 1 Mechanical properties of the base metal (σ_y: yield stress, σ_u: tensile strength)

Material	Thickness (mm)	σ_y (N/mm^2)	σ_u (N/mm^2)	Breaking elongation (%)
SN400B	9	284	435	32.3
	12	358	448	28.4
SS400	9	288	431	33.0
	12	356	438	27.0

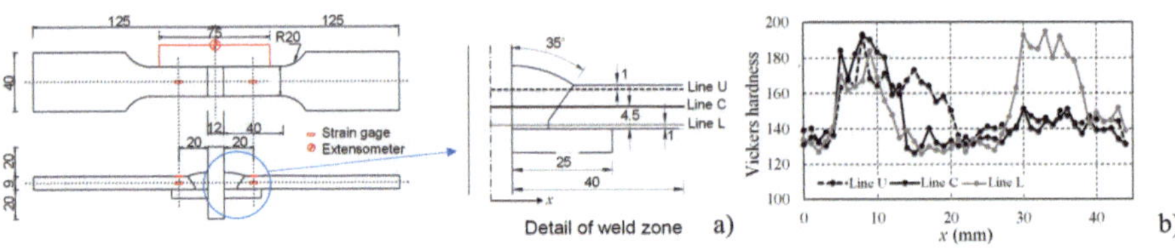

Fig. 6 Test specimen (SN400B, 9 mm thickness): (**a**) dimensions and (**b**) Vickers hardness

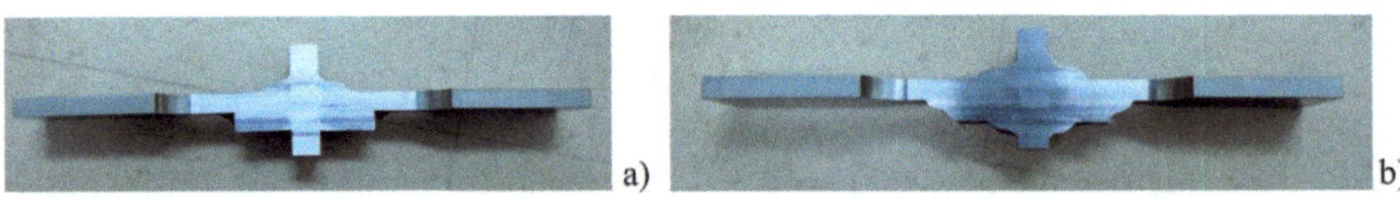

Fig. 7 Pictures of test specimens with and without tack welding (SN400B, 9 mm thickness): (**a**) specimen without tack welding and (**b**) specimen with tack welding

Vickers hardness along Lines U, C and L (see Fig. 6a). It is found that the weld metal is harder than the base metal. The nominal elastic modulus E is 2.05×10^5 N/mm^2. Figure 7 shows the pictures of test specimens.

5.2 *Test Procedure*

Fatigue tests are conducted using a 100 kN servo hydraulic fatigue testing machine (Fig. 8). Displacement control method using the output of an extensometer mounted at the center of the specimen with a gauge length of 75 mm is employed. The period of cyclic loading ranges from 0.5 s to 2 s. Strain gauges with a gauge length of 1 mm are attached to the specimen to measure the strains near the weld toe (see Fig. 6a). In addition to alternating strain tests (Test 1), in which the amplitude of nominal strain (ε_A) is varied over a wide range from 0.05% to 0.5%, cyclic loading tests with various mean strain levels ($\overline{\varepsilon}$) and strain amplitudes (ε_A) (Test 2) are performed to investigate the effects of mean strain on the fatigue strength. The combinations of $\overline{\varepsilon}$ and ε_A are as follows: (i) $\overline{\varepsilon} = 0.05\%$ and $\varepsilon_A = 0.05\%$, (ii) $\overline{\varepsilon} = 0.1\%$ and $\varepsilon_A = 0.05\%$, (iii) $\overline{\varepsilon} = 0.1\%$ and $\varepsilon_A = 0.1\%$, and (iv) $\overline{\varepsilon} = 0.2\%$ and $\varepsilon_A = 0.1\%$. A combination of alternating strain tests with two different amplitudes, ε_{A1} and ε_{A2}, (Test 3) is also conducted to investigate the validity of the Miner's law (the linear cumulative

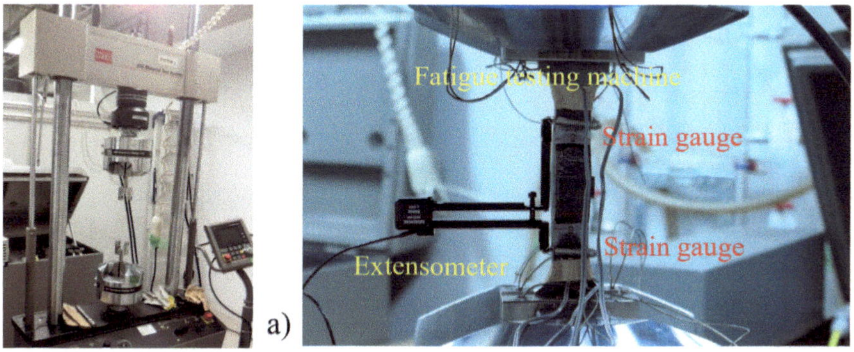

Fig. 8 Fatigue test: (**a**) fatigue testing machine and (**b**) test specimen and instrument

damage rule). The combinations of ε_{A1} and ε_{A2} are as follows: (i) $\varepsilon_{A1} = 0.05\%$ and $\varepsilon_{A2} = 0.1\%$ (Cases 1 and 2), (ii) $\varepsilon_{A1} = 0.1\%$ and $\varepsilon_{A2} = 0.05\%$ (Cases 3 and 4), (iii) $\varepsilon_{A1} = 0.15\%$ and $\varepsilon_{A2} = 0.2\%$ (Cases 5 and 6), and (iv) $\varepsilon_{A1} = 0.2\%$ and $\varepsilon_{A2} = 0.15\%$ (Cases 7 and 8). The specimen is first subjected to alternating strain of ε_{A1} $N_1/2$ times (Cases 1, 3, 5 and 7) or $N_1/4$ times (Cases 2, 4, 6 and 8), with N_1 being the number of cycles to fatigue fracture when the specimen is subjected to alternating strain of ε_{A1}. Then, the specimen is subjected to alternating strain of ε_{A2} until the specimen fails.

5.3 *Test Results*

Figure 9 shows the variation of the maximum nominal stress and the crack length with the number of loading cycles *N*, which is obtained from Test 1 with $\varepsilon_A = 0.9\%$. In the figure, the stress-strain hysteresis loops at the beginning of loading and just before the fracture are also illustrated. The nominal strain was measured by the extensometer. The crack length was analytically determined based on the maximum load-bearing capacity and the crack propagation path which was detected from the video images taken from two directions. The crack was initiated at the weld toe or root and propagated through the base metal or weld metal or along the bond. Then, the specimen finally failed. More than 80% of failure was initiated at the weld root.

The load-bearing capacity slightly decreases with the number of loading cycles *N*. As *N* approaches a critical number N_f, which is called the 'number of fracture cycles', the stiffness decreases and the crack length increases rapidly, leading to fracture. The number of loading cycles N_c when the load-bearing capacity decreases by 5% from the initial value is defined as the 'number of crack initiation cycles'. Hereafter, the number of crack initiation cycles is abbreviated as 'SN9_Nc' for SN400B specimens of 9 mm thickness, for example.

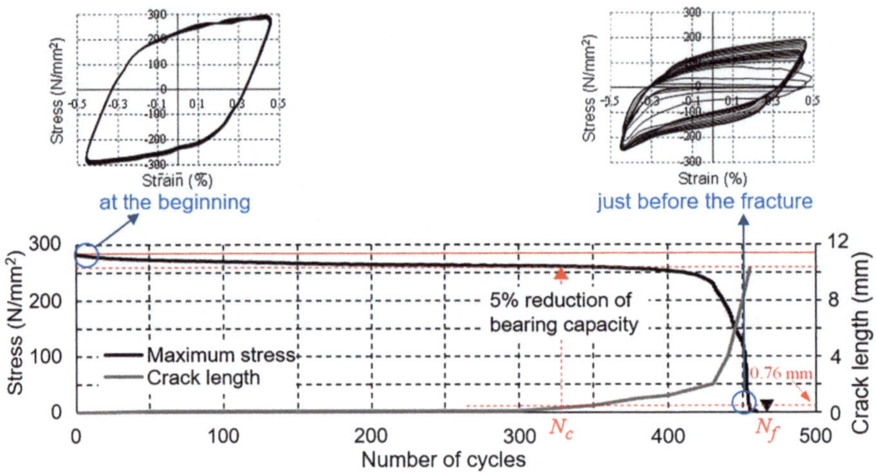

Fig. 9 Results of alternating strain test with $\varepsilon_A = 0.9\%$ (SN400B, 9 mm thickness)

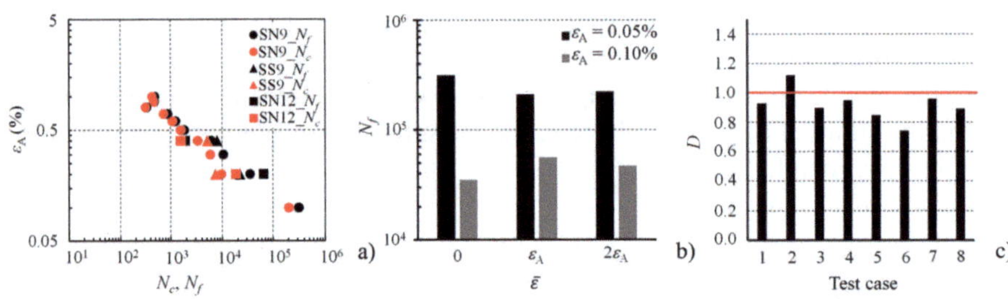

Fig. 10 Results of fatigue test (SN400B, 9 mm thickness): (**a**) fatigue curve (Test 1), (**b**) effects of $\bar{\varepsilon}$ on N_f and (**c**) degree of cumulative fatigue damage

Figure 10a shows the results of alternating strain tests (Test 1). When N_c and N_f are plotted against ε_A on a log-scale, the data points lie on straight lines. The effect of mean strain $\bar{\varepsilon}$ on N_f, which was obtained from Test 2, is shown in Fig. 10b. It is found that fatigue life is mainly determined by the strain amplitude ε_A and the effect of $\bar{\varepsilon}$ is relatively small within the limits of the present experiment. Figure 10c shows the degree of cumulative fatigue damage D, based on the Miner's law for each of eight test cases obtained from Test 3. The value of D is evaluated by the following equation:

$$D = \sum_j \left(n_j / N_{fj}\right) \tag{3}$$

where n_j = number of loading cycles for a given strain amplitude ε_{Aj}, and N_{fj} = number of loading cycles to fatigue fracture (fracture cycles) in the alternating strain test with ε_{Aj}, which can be obtained from Fig. 10a. According to the Miner's law, fatigue fracture occurs when D reaches 1. In Fig. 10c, the D value ranges from about 0.8 to 1.1. This feature implies that the Miner's law can be used for evaluating the fatigue life.

6 Evaluation of Fatigue Damage to the Welded Joints Using the Local Strain at Crack Initiation Location

6.1 FE Model

The magnitude and distribution of strains produced in the test specimen depends on the specimen's geometry, particularly on the shape of weld. Furthermore, it is thought that the fatigue strength is strongly dependent on the local strain at crack initiation location, which cannot be detected from the fatigue test. Thus, a finite element (FE) elasto-plastic analysis reproducing the above-mentioned fatigue test is employed.

The analysis is conducted using a commercial computer software 'ABAQUS 6.14.' First, the shape of the specimen is measured by a 3D scanner with an accuracy of 10 μm (Fig. 11a). Then, a 2D analytical model is created for the central cross-section of the specimen using the measured data. The FE model consists of 4-node shell elements, as shown in Fig. 11b. It reproduces only one half of the specimen considering the symmetry of the specimen. The shape of weld toe is modelled by a circle of radius ρ. The angular deformation ϕ is also determined, which induces initial stresses in the specimen when the specimen is set to the testing machine. ABAQUS CAE is used for the element decomposition. Finer meshes are used near the root and toe of weld. The smallest size is about 50 μm. The specimen is divided into three zones, i.e., base metal, weld metal and heat-affected zones. The mechanical properties in each zone are determined considering the effect of welding on the material in each zone. For the base metal zone, they are determined based on the result of a static tensile test. For the weld metal and heat-affected zones, they are determined based on the Vickers hardness [15]; the heat-affected zone is assumed to range from the bond to 3 mm.

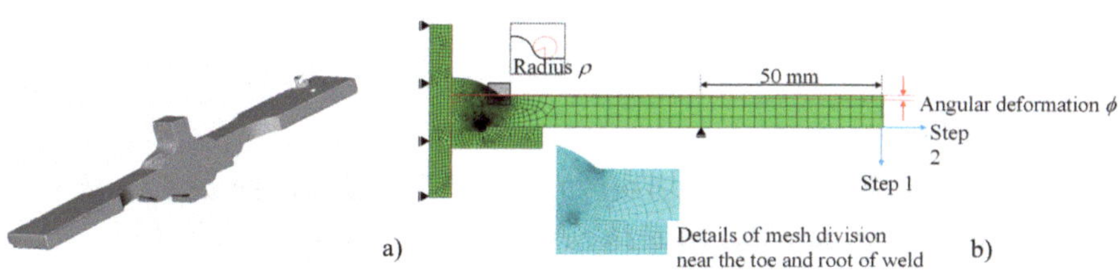

Fig. 11 Experimental and analytical models: (**a**) experimental model and (**b**) analytical model

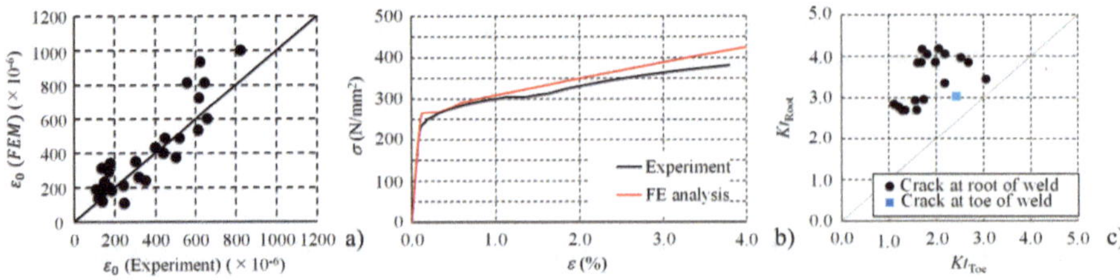

Fig. 12 Analysis results of the uniform tensile test for the test specimen: (**a**) initial strain ε_0, (**b**) σ—ε curve (SN400B, 9 mm) and (**c**) stress concentration factor

6.2 FE Analytical Results

The analysis is made for all specimens used in the fatigue test. It consists of two steps. First, the specimen is subjected to a forced displacement of 50 mm (jig fixing length) times tanϕ at the end (Step 1). Fig. 12a shows a comparison between experiment and FE analysis for the initial strains ε_0 at the strain gauge location near the toe of weld (see Fig. 6a). The results for all specimens are plotted in the figure. It is found that the agreement between experiment and FE analysis is generally good. Then, tensile load is applied to the specimen monotonically (Step 2). Figure 12b shows a comparison between experiment and FE analysis for the stress (σ) - strain (ε) curve obtained from the static tensile test with a SN400B specimen of 9 mm thickness. The experimental and analytical results correspond well to each other, indicating the validity of the analytical model and procedure used in the present study. Then, the crack length at each step of loading cycles is computed. In the analysis, the crack is reproduced by providing a slit along the crack propagation path observed in the experiment. As mentioned above, the crack is generally initiated at the toe or root of weld. The stress concentration factors *Kt*, defined by the ratio of local stress at crack initiation location to nominal stress, are computed using the stress distribution in the specimen obtained from the FE elastic analysis. Figure 12c shows the results for *Kt*. The cracks are initiated at the toe of weld (circle) or at the root of weld (square). The values of *Kt* are generally larger at the root of weld than at the toe of weld ($Kt_{\mathrm{Root}} > Kt_{\mathrm{Toe}}$). It is said that the cracks are generally initiated at the root of weld. In the beam-column welded joints where many roots and toes of weld exist, the stress concentration factor may play a significant role in detecting the crack initiation locations.

6.3 Relation Between the Amplitude of Local Strain at Crack Initiation Location and the Number of Loading Cycles to Fatigue Fracture

Figure 13a shows the relation between the strain amplitude and the number of fracture cycle N_f, in which both the amplitude of nominal strain (ε_A) and the amplitude of local strain at crack initiation location (ε_A^L) are considered. Similarly,

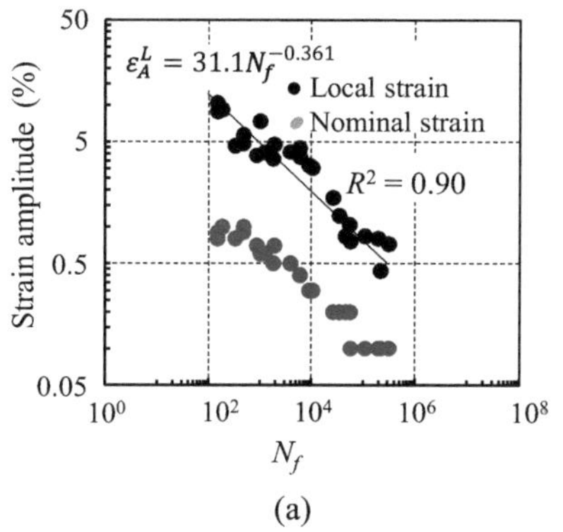

Fig. 13 Fatigue curve for beam-column welded joints: (**a**) number of loading cycles to fracture and (**b**) Number of loading cycles to crack

Figure13b shows the relation between the strain amplitude and the number of crack initiation cycle N_c. It is found that the values of ε_A^L are one order of magnitude larger than those of ε_A for both N_f and N_c. The relation between ε_A^L and N_f is approximated by the following equation:

$$\varepsilon_A^L = 31.1N_f^{-0.361} \tag{4}$$

Similarly, the relation between ε_A^L and N_c is approximated by the following equation:

$$\varepsilon_A^L = 80.3N_c^{-0.458} \tag{5}$$

Equations 4 and 5 can be used for evaluating the fatigue strength of beam-column welded joints of low-rise steel buildings more reasonably than the $\varepsilon_A - N_f$ and $\varepsilon_A - N_c$ curves, considering that the local strain at the crack initiation location significantly affects the fatigue strength and fatigue life and that ε_A depends on the shape of weld significantly. It can be said that the use of ε_A^L makes it possible to universally evaluate the fatigue strength of beam-column welded joints.

7 Evaluation of Wind-Induced Fatigue Damage to Welded Joints

The degree of cumulative fatigue damage (D) to the beam-column welded joints due to winds during a period of 10 minutes at various values of '*Basic wind speed*' U_0 defined in the AIJ-RLB [13] are computed in the time domain using the time histories of wind pressure coefficients $C_{pe,j}(t)$, the rainflow counting method and

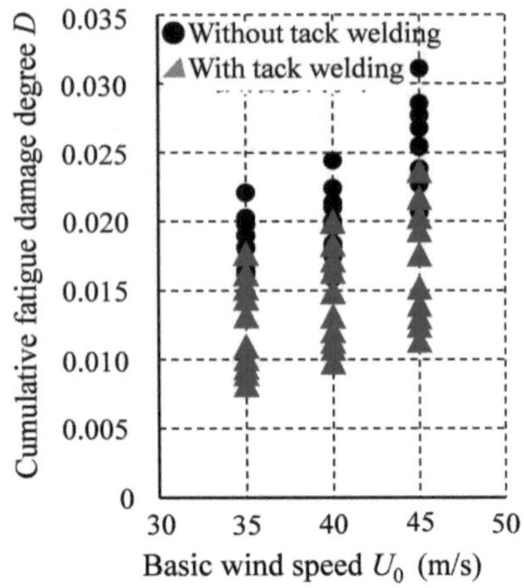

Fig. 14 Variation of D with '*Basic wind speed*' U_0 defined in AIJ-RLB [13]

the $\varepsilon_A^L - N_f$ curve (Fig. 13a and Eq. 4). The effect of mean stress is considered by using the Soderberg diagram [3]. Figure 14 shows the computed results, in which the D values for various frames and wind directions are plotted together. Therefore, the data plots scatter over a wide range. The D value for $U_0 = 35$ m/s (design wind speed) calculated by the method given in the JSSC Fatigue Design Recommendations for Steel Structures [3], in which the stress levels are assumed to be in the elastic range, was $D_{JSSC} = 2.14 \times 10^{-4}$. The maximum value of D for $U_0 = 35$ m/s shown in Fig. 14 is about 100 times the value of D_{JSSC}. This means that the degree of cumulative fatigue damage will be underestimated remarkably if the local plasticization of the weld joints is not considered appropriately. With an increase in U_0, the D value increases significantly.

8 Concluding Remarks

The present paper investigated the fundamental characteristics of wind-induced fatigue damage to the beam-column welded joints of flat-roofed low-rise steel buildings under practical conditions of high winds based on a wind tunnel experiment, a fatigue test and a finite element elasto-plastic analysis. In the wind tunnel experiment, we obtained the time histories of wind pressure coefficients at many points on the frames of the building in a turbulent boundary layer simulating natural winds over typical suburban exposure. The time histories of wind pressure coefficients were used for the fatigue damage analysis in the time domain. In the fatigue test, we obtained the relation between the amplitude of nominal strain ε_A and the number of loading cycles to fatigue fracture N_f (fatigue curve), which considered both low-cycle fatigue and high-cycle fatigue. Furthermore, the tensile test was reproduced by a finite element elasto-plastic analysis and the local strain at crack initiation location was obtained. Using the results of the fatigue test and finite element analysis, we obtained the relation between the amplitude of local strain at crack initiation location (ε_A^L) and N_f.

The main findings obtained from the present study may be summarized as follows:

1. The maximum extreme fiber stress occurs at the windward beam-column joint of an inside frame (Frame 2) at a wind direction of $\theta = 90°$ (parallel to the frame).
2. The strains induced in the welded joints may exceed the yield strain instantaneously even at the damage-limit design wind speed, considering that the yield strain of the weld metal is lower than that of the base metal. Thus, not only high-cycle fatigue but also low-cycle fatigue should be considered appropriately in the fatigue damage analysis of the frames.
3. Cracks may occur at the root or toe of weld.
4. The crack initiation location can be detected from the distribution of stress concentration factors obtained from a finite element elastic analysis.
5. The local strain at the crack initiation location may become about 10 times as large as the nominal strain. The relation between ε_A^L and N_f is approximated by Eq. (4).
6. The degree of cumulative fatigue damage D at the welded joint, which is predicted by the Miner's law with an assumption that the stress is within the elastic range, is about 1/100 as small as that evaluated by using the $\varepsilon_A^L - N_f$ curve considering the plastic behavior. Therefore, the local elasto-plastic behavior under fluctuating wind loading should be considered for accurately estimating the degree of accumulative fatigue damage. The values of ε_A^L can be evaluated by a finite element elasto-plastic analysis. However, regarding the material properties and the geometry of the welded joint to be used in the FE analysis, further investigations are required.

Acknowledgement The present study is financially supported by The Japan Iron and Steel Federation and The Japan Welding Engineering Society. The authors thank Prof. Seiichiro Tsutsumi of Osaka University for helpful suggestions.

References

1. The Japan Iron and Steel Federation: https://www.jisf.or.jp/business/tech/build/documents/
2. Repetto, M.P., Solari, G.: Dynamic crosswind fatigue of slender vertical structures. Wind Struct. **5**(6), 527–542 (2002)
3. Repetto, M.P., Solari, G.: Closed form solution of the alongwind-induced fatigue damage to structures. Eng. Struct. **31**, 2414–2425 (2009)
4. Japanese Society of Steel Construction: Fatigue Design Recommendation for Steel Structures (2012). (in Japanese)
5. Cheng, B., Qian, Q., Zhao, X.: Stress concentration factors and fatigue behavior of square bird-beak SHS T-joints under out-of-plane bending. Eng. Struct. **99**, 677–684 (2015)
6. Chen, K., Huang, H., Wu, Q., Nakamura, S., Chen, B.: Experimental and finite element analysis research on the fatigue performance of CHS K-joints. Eng. Struct. **197**, 109365 (2019)

7. Karabulut, B., Rossi, B.: On the fatigue behavior of duplex and high-strength welded cruciform joints. Eng. Struct. **247**, 113161 (2021)
8. Narihara, H., Izumi, M., Asami, Y.: Wind induced fatigue design for high rise steel buildings. J. Struct. Construct. Eng. AIJ. **465**, 129–137 (1994) (in Japanese)
9. Tanuma, Y., Fukasawa, T., Kobayashi, H.: Study on fatigue characteristics of cross welded joints of high strength steel. J. Struct. Construct. Eng. AIJ. **531**, 149–156 (2000) (in Japanese)
10. Ogawa, H., Ohkuma, T., Nakagomi, T.: Fatigue strength under uniform cycle load of large scale models—evaluation of fatigue failures at beam to column welded connections in steel structures part 1. J. Struct. Construct. Eng. AIJ. **535**, 149–156 (2000) (in Japanese)
11. Information Center for Building Administration, Japan Building Disaster Prevention Association: Commentary on structural regulations of the building standard law of Japan (2020). (in Japanese)
12. Japanese Industrial Standards Committee: JIS G 3136, Rolled Steels for Building Structures (2012)
13. Japanese Industrial Standards Committee: JIS G 3101, Rolled steels for general structures (2020)
14. Architectural Institute of Japan: Recommendations for Loads on Buildings (2015)
15. American Society of Civil Engineers: Wind Tunnel Testing for Buildings and Other Structures (2012)
16. American Society of Testing and Materials: ASTEM E140-12B Standard hardness conversion tables for metals relationship among Brinell Hardness, Vickers hardness, Rockwell hardness, Superficial hardness, Knoop hardness, scleroscope hardness, and leeb hardness (2019)

Torsional-Flutter-Based Harvester: Output Power in Nonstationary, Thunderstorm-Like Winds

Luca Caracoglia

Abstract This study continues the recent investigations on a flutter-based energy harvester. The work quantifies the output power of the harvesting device in non-ideal wind conditions. The numerical model replicates gusty and nonstationary features of a moderate-intensity thunderstorm outflow. A stochastic calculus differential equation is used to numerically examine the system performance.

Keywords Thunderstorm-like flows · Energy harvesting · Torsional flutter · Output power

1 Introduction

Wind-based energy harvesters, exploiting bending-torsion "classical" flutter instability have been examined in the literature (e.g., [10, 11]). Less attention has been paid to torsional-flutter-based energy conversion. Although torsional flutter usually entails low-performing conversion compared to classical flutter, a purely rotational mechanism about a pivot axis is advantageous since harvesting is achieved by means of a simple apparatus. The apparatus (Fig. 1) is composed of a rigid blade-airfoil of chord-wise length $2b$ ($0 < b \leq 1$ m) and span-wise length ℓ. The blade-airfoil is a NACA0012 streamlined profile of aspect ratio $AR = \ell/b = 4$. The blade-airfoil is attached to a support mast using a flexible, thin plate-like interface component, shown as "torsional-rotational" mechanism in Fig. 1. This interface is flexible and

L. Caracoglia (✉)
Visiting Professor (June 2024), Department of Civil & Environmental Engineering (DICA), University of Perugia, Perugia, Italy

Visiting Professor (January–May 2024), Department of Civil, Environmental & Mechanical Engineering (DICAM), University of Trento, Trento, Italy

Department of Civil and Environmental Engineering, 400 Snell Engineering Center, Northeastern University, Boston, MA, USA
e-mail: l.caracoglia@northeastern.edu

I. Calotescu, C. I. Coşoiu (eds.), *Proceedings of the Third National Conference on Wind Engineering*, Springer Proceedings in Earth and Environmental Sciences,
https://doi.org/10.1007/978-3-032-09671-5_2

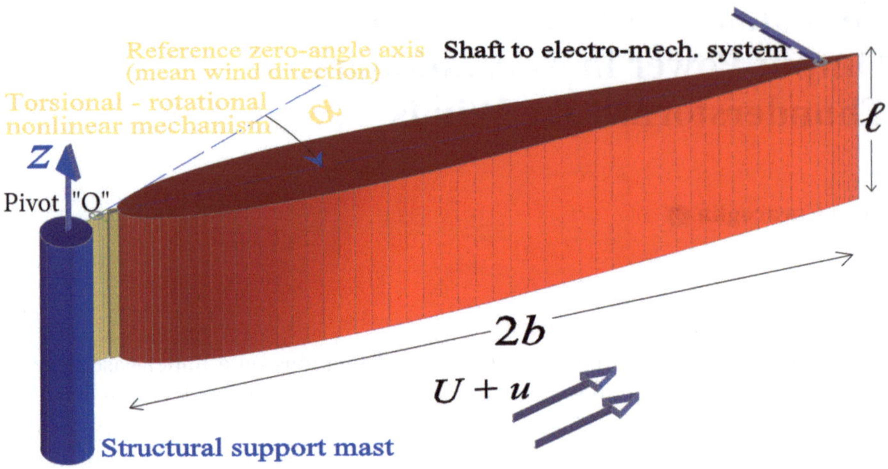

Fig. 1 Torsional flutter harvester: three-dimensional schematics

allows for the transmission of a nonlinear, elastic torsional restoring moment about pivot "O".

Recent work [3, 4] has explored the performance of the harvester in non-ideal winds, e.g., nonstationary gust fronts originating, for example, from thunderstorms. The work continues herein by examining the output power in various thunderstorm outflows as a function of geometry and dynamic properties.

Non-synoptic, nonstationary outflows specifically replicate non-ideal wind conditions, e.g., gust fronts associated with a thunderstorm downburst [13–15], as observed in urban settings [1], where the apparatus is planned to be deployed. The apparatus (Fig. 1) might possibly be installed on building roofs.

2 Modeling Background

2.1 Dynamic Flapping Model

The kinetic energy of the flow is converted to electrical energy by eddy currents, induced by a translating magnet inside a coil, not shown in Fig. 1 but described in previous works [2]. The rotation α is about the vertical axis z, i.e., the flapping occurs on a horizontal plane about the pivot "O" at the leading edge of the harvester's blade airfoil. The cross section of the blade-airfoil is streamlined, described by the NACA0012 profile. The blade-airfoil has one degree of freedom about "O", i.e., the rotation α.

The equilibrium equation reads, in dimensionless form:

$$\frac{d^2\alpha}{d\tau^2} + 2\zeta_\alpha \frac{d\alpha}{d\tau} + \alpha + \mathrm{H_{NL}}\left(\alpha, \frac{d\alpha}{d\tau}\right) = \frac{M_{0z} + M_{(\mathrm{e.m.})}}{\omega_\alpha^2 I_{0\alpha}} \tag{1}$$

In the previous equation the derivative $d\alpha/d\tau$ is carried out with respect to dimensionless time $\tau = \omega_\alpha t$ with dimensional t time; ω_α is the angular frequency of the linear torsional flapping oscillating foil and $I_{0\alpha}$ is the total polar mass moment of inertia of the flapping foil. Linear damping is simulated through damping ratio $0 < \zeta_\alpha < 1$.

In Eq. (1) the function $\mathrm{H_{NL}} = \left(\kappa\alpha^3 - 2\zeta_\alpha\gamma\alpha^2\right)$ is used to simulate the nonlinear part of the restoring torque mechanism, with $\kappa > 0$ and $\gamma > 0$ being two dimensionless constant parameters [3]. In parallel with the linear restoring torque about "O" in Eq. (1), the nonlinear restoring torque mechanism is either replicated by a Duffing model with polynomial law $\mathrm{H_{NL}} = \kappa\alpha^3$, depending on the flapping angle only, or a hybrid Duffing—van der Pol mechanism that also utilizes a negative, nonlinear (i.e., unstable) damping term $\mathrm{H_{NL}} = \left(\kappa\alpha^3 - 2\zeta_\alpha\gamma\alpha^2\right)$ (with $\gamma > 0$) to enhance the harvesting.

The aeroelastic torque M_{0z} is about pivot O in Fig. 1; modeling of M_{0z} is described in previous works [2]. Furthermore, noting that the moment arm is $2b$ from the pivot O to the leeward edge, the electro-motive torque is $M_{(\mathrm{e.m.})} = -(2b)\Phi_{(\mathrm{e.m.c.})} I(\tau)$, in which $I(\tau)$ is the output current of the power system, $\iota(\tau) = I(\tau)\left[2b\omega_\alpha\Phi_{(\mathrm{e.m.c.})}/R_C\right]^{-1}$ is the dimensionless output current, R_C is the resistance of the power circuit and $\Phi_{(\mathrm{e.m.c.})}$ is the dimensional electro-mechanical coupling coefficient [newton/ampère] [2].

2.2 *Nonstationary Wind Field Properties*

In stationary boundary layer flows, the mean wind speed U that triggers flutter is about $10 < U < 20$ m/s; flutter may be influenced by along-wind turbulence $u(\tau)$. In the case of nonstationary flow fields, there is a slowly-varying (mean) wind speed that varies with time $\tau = \omega_\alpha t$ (dimensionless): $U(\tau) = \Pi(\tau)U_{\max}$, where $0 \leq \Pi(\tau) \leq 1$ is the intensification/decay function and $U_{\max}$ is the maximum velocity of the nonstationary flow.

The main transitory feature of the nonstationary wind field, simulating the passage of a thunderstorm or a gust front, is the temporal intensification, $\Pi(\tau)$. The intensification (intensity factor) is derived from previous work [9]. If the mean flow speed is substituted by $\Pi(\tau) = U(\tau)/U_{\max}$, as explained in [4], we get:

$$\Pi(\tau) = \begin{cases} \frac{\tau}{\tau_{\max}} & 0 \leq \frac{\tau}{\tau_{\max}} \leq 1 \\ e^{-0.52\left[\frac{\tau}{\tau_{\max}} - 1\right]} & \frac{\tau}{\tau_{\max}} > 1 \end{cases} \tag{2}$$

In Eq. (2), $t_{\mathtt{max}}$ (or $\tau_{\mathtt{max}}$ in dimensionless units) is the time instant corresponding to maximum intensification of the nonstationary wind event. Quantity $t_{\mathtt{max}}/T \approx 0.52$, uses $T = 1200\,\mathrm{s}$ as the typical total thunderstorm duration from initial touchdown to full decay, and $t_{\mathtt{max}} \approx 630\,\mathrm{s}$ is chosen. Using typical values of ω_α, $\tau_{\mathtt{max}}$ is $400 < \tau_{\mathtt{max}} < 1000$ [4] if the total duration of the resulting, nonstationary wind event is compatible with the "Andrews Air Force Base" thunderstorm [9], or $200 < \tau_{\mathtt{max}} < 500$ if $t_{\mathtt{max}} \approx 300\,\mathrm{s}$ is extrapolated from data presented by Solari [12].

In addition, nonstationary ("ns") turbulence is described as $u_{\mathtt{ns}}(\tau) = \varphi_{\mathtt{mf}}(\tau)u(\tau)$, where $u(\tau)$ is a stationary random turbulence process and $0 < \varphi_{\mathtt{mf}}(\tau) \leq 1$ is a uniform modulation function ("mf") [12].

It is noted that $\tau_{\mathtt{max}}$ plausibly designates the dimensionless time instant of maximum turbulence intensification, the same quantity as $\tau_{\mathtt{max}}$ in Eq. (2). Therefore, $\varphi_{\mathtt{mf}}$ is found as [4]:

$$\varphi_{\mathtt{mf}}(\tau) = \mathrm{e}^{\left(\frac{\tau_{\mathtt{max}}-\tau}{\omega_\alpha}\right)} \left(\frac{\tau}{\tau_{\mathtt{max}}}\right)^{\tau_{\mathtt{max}}/\omega_\alpha} \tag{3}$$

Equation (3) is valid if $\tau_{\mathtt{max}} > 0$. As previously indicated, $\tau_{\mathtt{max}} = t_{\mathtt{max}}\omega_\alpha$ depends on the total duration of the nonstationary wind event. Furthermore, since the reference dimension of the blade-airfoil (proportional to $\sqrt{2b\ell}$) is small compared to the integral turbulence length scales, the nonstationary gusty field is coherent across the three-dimensional blade-airfoil of the apparatus. Finally, stationary turbulence is normalized as $\hat{u}(\tau) = u(\tau)/U_{\mathtt{max}}$ and it is approximated by a Gaussian white noise, acceptable from the perspective of output power estimation [4].

2.3 Stochastic Differential Equation Model

Since the turbulence variable $\hat{u}$ is expressed by a Gaussian process, $\hat{u}$ is represented by a Wiener process of unit variance Gaussian increments and re-cast into a vector stochastic differential equation, Itô-type [5]:

$$\mathrm{d}\mathbf{W}_{\mathtt{em}} = \mathbf{q}_{\mathrm{em,NL}}\left(\mathbf{W}_{\mathrm{em}}\right)\mathrm{d}\tau + \sqrt{2\pi}\left[\mathbf{t}_{\mathrm{NL},\hat{u}}\left(\mathbf{W}_{\mathrm{em}}\right)\mathrm{d}B_{\hat{u}}(\tau) + \mathbf{Q}_{\mathrm{L},\Delta 2}\mathbf{W}_{\mathrm{em}}\mathrm{d}B_{\Delta 2}(\tau)\right] \tag{4}$$

$\mathbf{W}_{\mathtt{em}} = \left[\alpha, \mathrm{d}\alpha/\mathrm{d}\tau, \nu_{ae,1}, \nu_{ae,2}, \mu_{ae,1}, \mu_{ae,2}, \iota\right]^T$ in Eq. (4) is the state vector; $\mathbf{W}_{\mathtt{em}}$ includes both physical, aeroelastic states ($\nu_{ae,1}, \nu_{ae,2}, \mu_{ae,1}, \mu_{ae,2}$) and dimensionless output current ι. In Eq. (4), the scalar, Wiener noise $B_{\hat{u}}(\tau)$ of unit variance increments separately addresses turbulence perturbation from the noise $B_{\Delta 2}(\tau)$, used for load perturbation [4]. Quantity $\mathbf{q}_{\mathrm{em,NL}}$ is a nonlinear vector-function; $\mathbf{t}_{\mathrm{NL},\hat{u}}$ is a nonlinear turbulence diffusion vector-function. $\mathbf{Q}_{\mathrm{L},\Delta 2}$ is a constant, diffusion matrix that controls the load perturbation and depends on the standard deviation of $\Delta_2(\tau)$, σ_{d2} (see Sect. 3). Both $\mathbf{q}_{\mathrm{em,NL}}$ and $\mathbf{t}_{\mathrm{NL},\hat{u}}$ incorporate dependency on $\Pi(\tau)$, $\varphi_{\mathtt{mf}}(\tau)$ and $\sigma_{\hat{u}}$. Derivations are omitted for the sake of brevity but may be found in [2, 4].

Equation (4) clearly depends on the maximum mean wind speed $U_{\mathtt{max}}$, which is also employed as an independent input parameter. As the input mean wind speed varies, several conditions are examined: stable dynamic response and no flutter, torsionally driven flapping at flutter, post-critical operational conditions, etc. [4].

Solution to Eq. (4) is found in "weak form" by numerical integration, using the Euler-Maruyama method [7]. A suitable sample of the state vector $\mathbf{W}_{\mathtt{em}}(\tau = 0)$ with appropriate initial conditions is synthetically generated and, utilizing each realization as an independent initial condition, Eq. (4) is solved multiple times. Each quantity of interest (e.g., output power as in Sect. 2.4) is subsequently derived by Monte-Carlo methods [3, 4].

2.4 Mean Output Power Estimation

The output power is a stochastic process because torsional flutter is undertaken under slowly varying $U(\tau)$ and nonstationary random $u_{\mathtt{ns}}(\tau)$ in Eq. (4). The output power expectation at sustained flapping is [4]:

$$\mathbb{E}\left[\mathrm{P}_{\mathtt{out}}(\tau)\right] = \omega_\alpha^3 \Psi I_{0\alpha} \mathbb{E}\left[\iota^2(\tau)\right], \tag{5}$$

where $\mathbb{E}[.]$ denotes expectation, $\mathbb{E}\left[\iota^2(\tau)\right]$ is the time-dependent variance of ι, noting that $\mathbb{E}[\iota] \approx 0$; Ψ is the "dimensionless electro-mechanical coupling coefficient":

$$\Psi = \frac{4b^2 (\Phi_{(\mathtt{e.m.c.})})^2}{\omega_\alpha I_{0\alpha} R_C} \tag{6}$$

Because of the nonstationary flow features, Eq. (5) is also time dependent. Although other quantities may be considered to describe the efficiency of the harvester (e.g., dimensionless output power coefficient, quality factor [3]), the use of a dimensional output power is preferable since Eq. (5) is a simple expression that is calculated numerically and it is dimensional [W, watts].

3 Description of the Units, Turbulence and Load Perturbations

The simulations examine the efficiency of an apparatus with flapping pivot axis at the apex of the blade-airfoil and AR $= 4$ ($\ell = 4b$).

Three reference configurations or "Types" are selected from previous studies [2]. The main properties are described in Table 1. In this table, quantities such as angular frequency ω_α and damping ratio ζ_α must be interpreted as the properties of the linearized dynamic equation of the harvester.

Table 1 Harvester geometry, wind turbulence and load properties

Type	b [m]	AR	$\frac{I_{0\alpha}}{\ell}$ $[\frac{\text{kgm}^2}{\text{m}}]$	$\frac{\omega_\alpha}{2\pi}$ [Hz]	ζ_α [%]	κ	γ	$\sigma_{\hat{u}}$ [%]	$\tau_{\max}$	σ_{d2}
"Ty.0"	0.25	4	20	0.25	0.25	100	{0,0.5,1}	{2,10,20}	{190, 400}	0.07
"Ty.1"	0.25	4	40	0.25	0.30	100	{0,0.5,1}	{2,10,20}	{190, 400}	0.07
"Ty.2"	0.50	4	300	0.10	0.30	100	{0,0.5,1}	{2,10,20}	{190, 400}	0.07

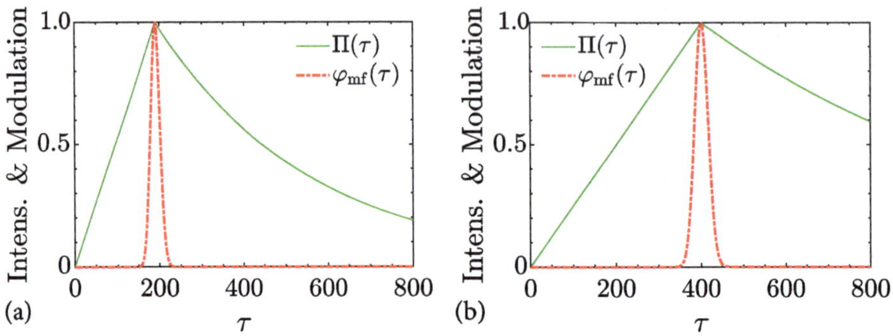

Fig. 2 Intensity and modulation functions of the nonstationary wind, thunderstorm-like outflows: (**a**) short ($\tau_{\max} = 190$), (**b**) long ($\tau_{\max} = 400$) duration

Furthermore, in Table 1 nonlinear stiffness parameter of the Duffing model is set to a constant $\kappa = 100$ in $\text{H}_{\text{NL}} = \kappa\alpha^3$, and with $\gamma = 0$ identically. By contrast, in the case of or the hybrid Duffing—van-der-Pol model [3] a variable nonlinear damping parameter $\gamma \geq 0$ is used with $\kappa = 100$ in $\text{H}_{\text{NL}} = \left(\kappa\alpha^3 - 2\zeta_\alpha\gamma\alpha^2\right)$.

In Table 1, the examined wind flow is nonstationary, with maximum mean speed $10 \leq U_{\max} \leq 40$ m/s, considering a wide range of options from relatively low to unrealistically high speed values. The turbulence intensity of the stationary part of the flow, $\hat{u}$, varies between 2% (smooth flow) and 20% (high turbulence). In Table 1, the temporal intensification time $\tau_{\max}$ is also varied to study the effects of the nonstationary outflow duration on the flutter onset; $\tau_{\max}$ varies between 190 and 400, i.e., the dimensional time duration of the nonstationary flow event approximately varies between 600 and 1200 s. The corresponding intensity and modulation functions of the wind outflows are depicted in Fig. 2. Finally, aeroelastic load perturbation is considered in Table 1. It is simulated, independently of turbulence effects, as a Gaussian white noise applied to the d_2 load parameter [2] of the Wagner's indicial function $\Phi(\tau) = [1 - c_1 e^{-d_1\tau k_\alpha^{-1}} - c_2 e^{-d_2\tau k_\alpha^{-1}}]$ [16],i.e., $d_2 = (d_{2,m} + \Delta_2(\tau))$ with mean value $d_{2,m} = 0.3$ [6] and standard deviation of the zero-mean noise $\Delta_2(\tau)$ equal to $\sigma_{d2} = 0.07$. This standard deviation corresponds to a coefficient of variation equal to about 23%. The reason for exclusively considering randomness in the d_2 parameter is because the second exponential term $\{c_2 e^{-d_2\tau k_\alpha^{-1}}\}$ primarily controls the unsteady load variations through the indicial function [4].

4 Output Power Results

4.1 Duffing Restoring Torque Mechanism

The output power of the Duffing-type apparatus, $\mathrm{P_{out}}(\tau)$, primarily depends on the dimensionless electro-mechanical coupling coefficient Ψ in Eq. (6). The expected value of $\mathrm{P_{out}}(\tau)$ is proportional to the mean-square value of the zero-mean, dimensionless random current $\iota(\tau)$, found by Faraday's Law [8] in Eq. (6).

Contrary to stationary wind flows characterized by constant mean wind speed and turbulence intensity [2], wind energy conversion is less efficient in nonstationary flows [4]. More specifically, such flows are unable to trigger flutter instability of the harvester and to induce a sustained post-critical flapping regime because of the variable wind conditions (Fig. 2). Furthermore, no relevant energy conversion is noted for maximum wind speeds $U_{\mathtt{max}}$ below 20–25 m/s and for smooth flow conditions (i.e., $\sigma_{\hat{u}} = 2\%$). Specific details are omitted for the sake of brevity but discussion may be found in a recent study [4].

As a first example, Fig. 3 illustrates two cases of mean output power *vs.* time τ, during a nonstationary, thunderstorm-like outflow of total duration 600 s, corresponding to about 400 in τ-units. The flows are characterized by moderate turbulence of intensity $\sigma_{\hat{u}} = 10\%$ (maximum value). The time instant $\tau_{\mathtt{max}} = 190$ corresponds to the maximum intensification of the nonstationary flow, as shown in Fig. 2a.

Two cases are presented, at $U_{\mathtt{max}} = \{14.4, 28.2\}$ m/s, and for various apparatus types (Table 1). Type-2 ("Ty.2") apparatus with linear angular frequency $\omega_\alpha = 0.628$ rad/s exhibits sustained flapping at $U_{\mathtt{max}} = 28.2$ m/s that is capable of producing about 0.02 W. In a previous study [4] $U_{\mathtt{max}} = 27.4$ m/s was indicated as the lower limit threshold for active, adequate conversion of the incoming flow energy. By contrast, no notable energy conversion is detected at lower speeds, i.e., $U_{\mathtt{max}} = 14.4$ m/ in Fig. 3a (note the different scale of the vertical axes). As the turbulence intensity of the nonstationary flow is increased to high values ($\sigma_{\hat{u}} = 20\%$), perhaps representative of urban flows contaminated by obstacles and buildings, the energy conversion mechanism is enhanced, with unstable flutter flapping triggered at lower mean wind speeds $U_{\mathtt{max}}$. Nevertheless, numerical integration issues of Eq. (4) are also observed, preventing the exact quantification of the output power at very large flow speeds beyond 30 m/s.

Figure 4 presents, as a second example, two panels (a, b) with the numerical time histories $\mathbb{E}\left[\mathrm{P_{out}}(\tau)\right]$ for Duffing-type apparatuses in non-ideal, short-duration nonstationary winds at $U_{\mathtt{max}} = \{20.0, 28.0\}$ m/s. Clearly, larger propensity to energy conversion is already possible at $U_{\mathtt{max}} = 20.0$ m/s in Fig. 4a with an instantaneous output power close to 0.03 for Type-2 ("Ty.2") apparatus, a promising value although insufficient. Moreover, simulations at $U_{\mathtt{max}} = \{20.0, 28.0\}$ m/s also confirm the peculiar nature of the nonstationary, thunderstorm—like outflow. A non-zero producible power is mainly concentrated in the interval $200 \leq \tau \leq 400$, in

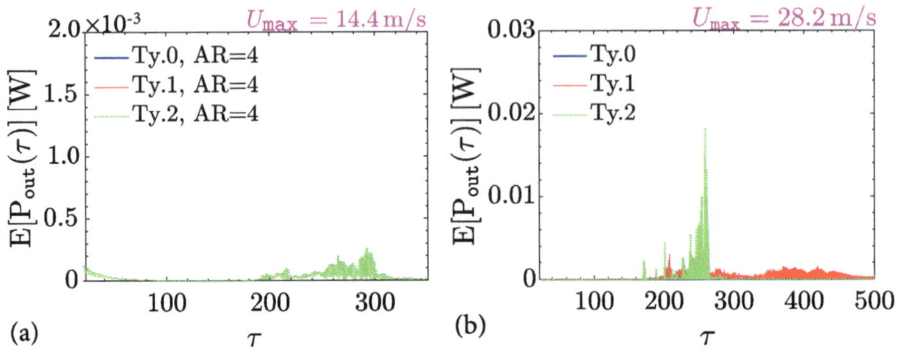

Fig. 3 Mean output power $\mathbb{E}\,[\mathrm{P_{out}}(\tau)]$ of Duffing-type apparatuses in non-ideal, short-duration ($\tau_{max} = 190$) nonstationary winds: maximum speed $U_{max} = \{14.4, 28.2\}$ m/s respectively (**a**) (**b**); moderately turbulent flows of intensity $\sigma_{\hat{u}} = 10\%$ [note: vertical axis scales are different in the figure panels]

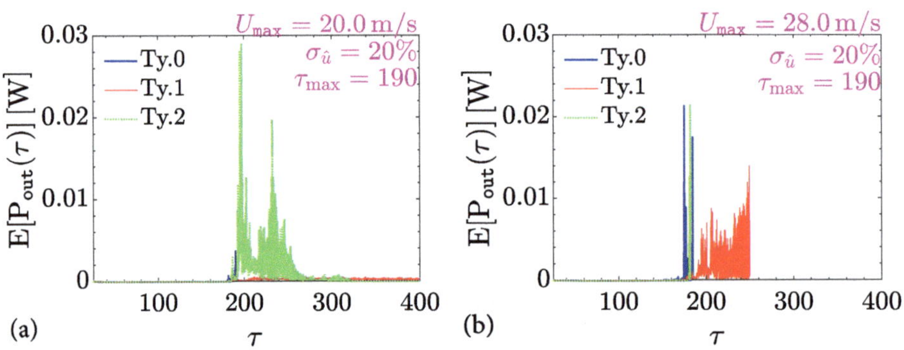

Fig. 4 Mean output power $\mathbb{E}\,[\mathrm{P_{out}}(\tau)]$ of Duffing-type apparatuses in non-ideal, short-duration ($\tau_{max} = 190$) nonstationary winds: maximum speed $U_{max} = \{20.0, 28.0\}$ m/s respectively (**a**), (**b**); high turbulence flows of intensity $\sigma_{\hat{u}} = 20\%$

which both intensification function $\Pi(\tau)$ and turbulence modulation functions $\varphi_{\mathtt{mf}}$ are considerably greater than zero (Fig. 2a).

Furthermore, Type-0 ("Ty.0") and Type-1 ("Ty.1") apparatuses are consistently less performing, with mean output power being consistently one order of magnitude lower than $\mathbb{E}\,[\mathrm{P_{out}}(\tau)]$ of Type-2 one in both Figs. 3 and 4. Preliminary, promising output power results are only found for Type-1 ("Ty.1") apparatus in Fig. 4b.

Finally, a longer duration wind event is examined in Fig. 5 for a Duffing-type apparatus. Although the total duration of the nonstationary event doubles, the active portion of the flow is primarily concentrated around the peak value $\tau_{max} = 400$ and, therefore, the non-negligible energy conversion is observed in the same τ temporal interval, which is influenced by turbulence modulation function $\varphi_{\mathtt{mf}}$ in Fig. 2b. In addition, simulations suggest that Type-1 apparatus can also be triggered in Fig. 5, exhibiting output powers equal to about 0.01 W at $\tau \approx 600$. Nevertheless, the

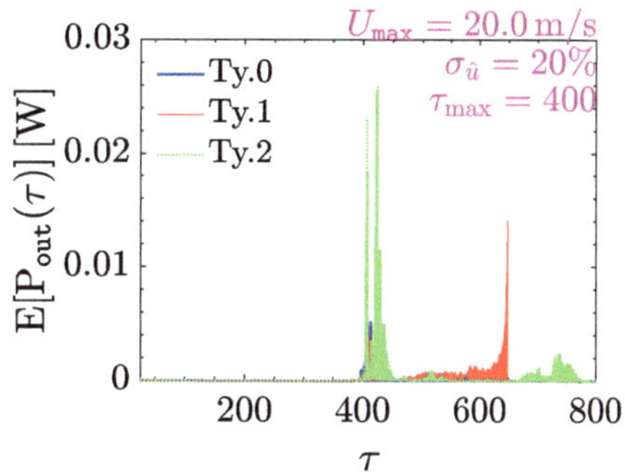

Fig. 5 Mean output power $\mathbb{E}[\mathrm{P_{out}}(\tau)]$ of Duffing-type apparatuses in non-ideal, long-duration ($\tau_{max} = 400$) nonstationary winds: maximum speed $U_{max} = 20.0$ m/s; high turbulence flows of intensity $\sigma_{\hat{u}} = 20\%$

numerical integration is interrupted for Type-1 apparatus beyond 650 in Fig. 5 since the nonlinear Duffing restoring torque generate a diverging time history pattern that may be numerically generated by the Euler-Maruyama numerical method [7], used to solve Eq. (4).

4.2 Hybrid Duffing: Van Der Pol Restoring Torque Mechanism

Previous, preliminary studies [3] suggested that a hybrid Duffing—van der Pol restoring torque mechanism may be moderately beneficial to the output power conversion. Results are extended in this subsection to consider the effects of variable nonlinear damping parameter γ, variable nonstationary wind field properties, i.e., duration of the event, turbulence intensity, etc. As observed with the Duffing - type mechanism, numerical integration issues are noticed for U_{max} beyond 25 m/s and high turbulence flows ($\sigma_{\hat{u}} = 20\%$). However, the general remark is that the hybrid restoring torque mechanism is beneficial, since non-negligible output powers $\mathbb{E}[\mathrm{P_{out}}(\tau)]$ are usually noted at lower maximum wind speed values U_{max} compared to Figs. 3 and 4.

For example, Fig. 6 presents the output power results at $U_{max} = 19.5$ m/s for a hybrid Duffing—van der Pol type apparatus with variable nonlinear damping: (a) $\gamma = 0.5$ and (b) $\gamma = 1.0$. This case refers to a high-turbulence, short-duration nonstationary thunderstorm-like wind flow. The mean output power $\mathbb{E}[\mathrm{P_{out}}(\tau)]$ evolves over time τ, compatibly with the intensification and turbulence modulations functions (Fig. 2a). Interestingly, the maximum value of the output power, equal to about 0.03 W for Type-2 apparatus, is obtained at a slightly lower mean wind speed ($U_{max} = 19.5$ m/s) compared to Fig. 4a ($U_{max} = 20.0$ m/s). Even though the improved performance is marginal in terms of absolute output power, the temporal duration of the active energy conversion stage with non-negligible $\mathbb{E}[\mathrm{P_{out}}(\tau)] > 0$ is prolonged.

Moreover, a variation of the van der Pol nonlinear damping properties in Fig. 6 [(a) $\gamma = 0.5$, and (b) $\gamma = 1.0$] does not lead to a significant change in the anticipated energy production. As already noted in [3], the flapping amplitudes are possibly low

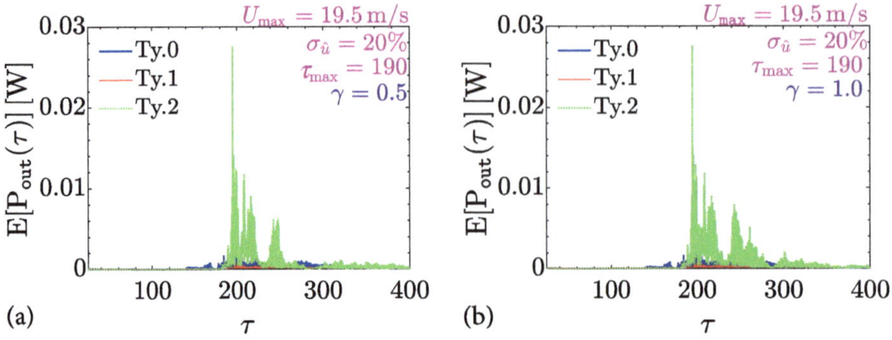

Fig. 6 Mean output power $\mathbb{E}[P_{out}(\tau)]$ of hybrid Duffing—van der Pol type apparatuses with variable nonlinear damping (**a**) $\gamma = 0.5$ and (**b**) $\gamma = 1.0$, in non-ideal, short-duration ($\tau_{max} = 190$) nonstationary winds at maximum speed $U_{max} = 19.5$ m/s and high turbulence flows ($\sigma_{\hat{u}} = 20\%$)

Fig. 7 Mean output power $\mathbb{E}[P_{out}(\tau)]$ of a hybrid Duffing—van der Pol type apparatus with nonlinear damping $\gamma = 0.5$, in non-ideal, long-duration ($\tau_{max} = 400$) nonstationary winds at maximum speed $U_{max} = 20.5$ m/s and moderate turbulence flows ($\sigma_{\hat{u}} = 10\%$)

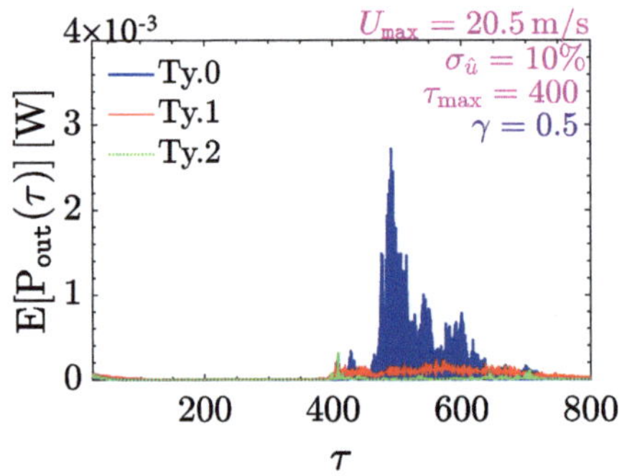

as to engage the negative nonlinear damping effect and induce a more unstable, i.e., operational regime for the harvester.

Long-duration nonstationary wind events with $\tau_{max} = 400$ and high turbulence flows ($\sigma_{\hat{u}} = 20\%$) tend to produce similar numerical results; the time histories of $\mathbb{E}[P_{out}(\tau)]$ are essentially quite similar to the panels in Fig. 6. Details are omitted for the sake of brevity. In any case, further examination of the energy production is considered in Fig. 7 for moderately turbulent, long-duration thunderstorm-like flows.

Figure 7 examines, as an example, the mean output power $\mathbb{E}[P_{out}(\tau)]$ of a hybrid Duffing—van der Pol type apparatus with nonlinear damping $\gamma = 0.5$, in non-ideal, moderately turbulent, long-duration ($\tau_{max} = 400$) nonstationary winds. Noting that the vertical axis scale is smaller compared to previous figure panels, the power converted by the harvester is rather small, almost imperceptible. Despite the longer exposure to wind flow the produced energy is not sufficient. This insufficient performance is plausibly associated with a combination of delayed intensification $\Pi(\tau)$ and lower turbulence intensity ($\sigma_{\hat{u}} = 10\%$).

5 Discussion and Conclusions

Numerical results demonstrate that it is possible to extract wind energy by means of the proposed harvester in non-ideal, transient flows. The main flow parameter that influence the power conversion are the maximum turbulence intensity, instantaneously varying with time. The total duration of the nonstationary, thunderstorm-like outflow appear to be irrelevant. In this paper, the above-described flow parameters were systematically varied, and $P_{out}(\tau)$ was determined.

Nevertheless, supplementary numerical analysis is needed to examine the effects of the specific flow conditions that could still enable power generation through the apparatus in nonstationary winds. Despite recent research advancements, there are still possibilities for improving the analysis and the performance of the harvester, e.g., through nonlinear aeroelastic load analysis, by CFD modeling, through experimental verification, etc. Future activity will focus on these important research items.

Acknowledgments L. Caracoglia gratefully acknowledges the support of the Universities of Trento and Perugia (Italy), where this study was completed during a sabbatical leave in 2024. This material is based in part upon work supported by the National Science Foundation (NSF) of the United States of America, Award CMMI-2020063. Any opinions, findings and conclusions or recommendations are those of the author and do not necessarily reflect the views of the NSF.

References

1. Burlando, M., Romanić, D., Solari, G., Hangan, H., Zhang, S.: Field data analysis and weather scenario of a downburst event in Livorno, Italy, on 1 October 2012. Mon. Weather Rev. **145**, 3507–3527 (2017)
2. Caracoglia, L.: Modeling the coupled electro-mechanical response of a torsional-flutter-based wind harvester with a focus on energy efficiency examination. J. Wind Eng. Ind. Aerodyn. **174**, 437–450 (2018)
3. Caracoglia, L.: Improving output power of a torsional-flutter harvester in stochastic thunderstorms by Duffing — van der Pol restoring torque. ASCE-ASME J. Risk U. B **10**(4), 041204 (2024)
4. Caracoglia, L.: Stochastic performance of a torsional-flutter harvester in non-stationary, turbulent thunderstorm outflows. J. Fluids Struct. **124**, 104050 (2024)
5. Itô, K.: On stochastic differential equations. Memoirs of the American Mathematical Society, vol. 4. American Mathematical Society, Providence (1951). ISBN: 978-0-8218-1204-4
6. Jones, R.T.: The unsteady lift of a finite wing. Technical Note 682, National Advisory Committee for Aeronautics, Washington, D.C., USA (1939)
7. Kloeden, P.E., Platen, E., Schurz, H.: Numerical Solution of Stochastic Differential Equations Through Computer Experiments. Springer-Verlag, Berlin-Heidelberg (1994)
8. Kwon, S.D., Park, J., Law, K.: Electromagnetic energy harvester with repulsively stacked multilayer magnets for low frequency vibrations. Smart Mater. Struct. **22**(5), 055007 (2013)
9. Le, T.H., Caracoglia, L.: Computer-based model for the transient dynamics of a tall building during digitally simulated Andrews AFB thunderstorm. Comput. Struct. **193**, 44–72 (2017)
10. Matsumoto, M., Mizuno, K., Okubo, K., , Ito, Y.: Fundamental study on the efficiency of power generation system by use of the flutter instability. In: Proceedings of the 2006 ASME Pressure Vessels and Piping Division Conference, vol. 9, pp. 277–286. American Society of Mechanical Engineers (2006)

11. Shimizu, E., Isogai, K., Obayashi, S.: Multiobjective design study of a flapping wind power generator. J. Fluids Eng.-ASME **130**(2), 021104 (2008)
12. Solari, G.: Characteristics of thunderstorms relevant to the wind loading of structures. Wind Struct. **20**(6), 763–791 (2015)
13. Solari, G.: Thunderstorm response spectrum technique: theory and applications. Eng. Struct. **108**, 28–46 (2016)
14. Solari, G., De Gaetano, P.: Dynamic response of structures to thunderstorm outflows: response spectrum technique vs time-domain analysis. Eng. Struct. **176**, 188–207 (2018)
15. Solari, G., De Gaetano, P., Repetto, M.P.: Thunderstorm response spectrum: fundamentals and case study. J. Wind Eng. Ind. Aerodyn. **143**, 62–77 (2015)
16. Wagner, H.: Über die entstehung des dynamischen auftriebes von tragflügeln (in German). ZAMM - Zeitschrift für Angewandte Mathematik und Mechanik **5**, 17–35 (1925)

Evaluating the Impact of Chamfer Corners on Building Aerodynamics Under Interference Effects Using LBM-LES CFD Simulation

Saiful Naim Sulaiman, Christian F. Janssen, and Thomas Indinger

Abstract The impact of chamfer corners for a tall building under the interference effect was studied using Large Eddy Simulation (LES), which was solved using the Lattice Boltzmann Method (LBM). Two buildings, identical in size, were spaced at 1.5 times their width. Six chamfer rates between 0% and 25% were applied on the principal building and tested under different wind incidence angles. The Aerodynamic Interference Factors (AIF) of base moment coefficients for along wind, across wind, and torsional components were evaluated to quantify the interference effect. It was found that increasing chamfer rates generally lower the AIF, with a 15% chamfer rate providing the optimal reduction across most conditions. However, a 20% chamfer rate produced decreased performance and even adverse effects, particularly at a 135° wind direction, where it increased the root-mean-square (RMS) across wind base moment coefficient, C'_{ML}. Furthermore, chamfer modifications were less effective against along wind forces, particularly when the shielding effect from the interference building is dominant. This study provides insights into the effectiveness of chamfering as an aerodynamic modification in urban settings to enhance structural stability and reduce wind-induced loads.

Keywords Wind-induced loads · Chamfered corners · Interference effect

S. N. Sulaiman (✉)
Chair of Aerodynamics and Fluid Mechanics, Technical University of Munich, Garching, Germany

Faculty of Mechanical Technology and Engineering, Universiti Teknikal Malaysia Melaka, Durian Tunggal, Malaysia
e-mail: saiful.sulaiman@tum.de

C. F. Janssen
Altair Engineering GmbH, Böblingen, Germany

T. Indinger
Chair of Aerodynamics and Fluid Mechanics, Technical University of Munich, Garching, Germany

I. Calotescu, C. I. Coşoiu (eds.), *Proceedings of the Third National Conference on Wind Engineering*, Springer Proceedings in Earth and Environmental Sciences,
https://doi.org/10.1007/978-3-032-09671-5_3

1 Introduction

Chamfering the corners of a square or rectangular building is one of the modification techniques to reduce aerodynamic loads induced by the incoming wind. This method alters the flow separation points, reduces vortex shedding, and minimizes dynamic wind loads, making it a crucial modification in the design of tall buildings. Numerous studies have investigated the impact of chamfered corners on aerodynamic performance. Kawai [1] found that small corner cuts and recessions increase aerodynamic damping, which helps prevent aeroelastic instabilities like vortex-induced vibration and galloping. However, large modifications can sometimes promote instability at lower wind speeds. Similarly, Tamura and Miyagi [2] reported that chamfered and rounded corners reduce drag forces by decreasing wake width, especially in turbulent flows where shear layers re-attach to the building's side surfaces, enhancing aerodynamic stability. Most studies suggest that a 10% chamfer rate is generally recommended for improving aerodynamic performance.

Experimental studies, such as those by Li et al. [3] have shown that chamfered and recessed corners reduce wind loads by altering flow patterns. Similarly, Yan and Li [4] demonstrated that chamfered corners mitigate wind interference, reducing aerodynamic forces and moments. These findings are further supported by numerical simulations using Computational Fluid Dynamics (CFD). For instance, Shang et al. [5] and Cao et al. [6] utilized CFD to show that chamfered corners reduce aerodynamic forces and interference, providing detailed insights into flow dynamics and pressure distributions. The trend of using Large Eddy Simulations (LES) in CFD is growing due to its high accuracy in capturing complex flow fields. LES can model the turbulent flows around buildings with greater precision than traditional methods. This accuracy is essential for understanding aerodynamic interactions with different chamfer rates. Studies employing LES, such as those by Cao et al. [6], have demonstrated the significant benefits of this method in providing detailed aerodynamic analysis.

In recent years, research has also moved toward the aerodynamic optimization of building shapes using advanced surrogate modeling techniques. For example, Ding and Kareem [7] introduced a multi-fidelity surrogate modeling framework combining RANS and LES simulations to automatically explore optimal geometries for wind load mitigation, offering a balance between computational efficiency and accuracy. Similarly, Elshaer et al. [8] proposed an aerodynamic optimization procedure (AOP) that couples LES with an artificial neural network (ANN) to minimize wind-induced forces on tall buildings. Their findings demonstrated that a two-surface chamfer constrained to 20% of the building width could reduce both along and across wind responses by over 30%, with minimal impact on architectural design. While these studies established important benchmarks for isolated buildings, their findings require validation in interference conditions typical of urban environments.

The Lattice Boltzmann Method (LBM) has emerged as an efficient alternative for LES implementations, as demonstrated by Adeeb, Haider, and Sohn [9], who achieved high accuracy in simulating complex flow patterns around buildings. These advances in simulation techniques have made possible the detailed investigation of interference effects that is central to the current study.

Most of the existing studies have focused on isolated or single buildings. However, there is growing recognition of the importance of investigating buildings under interference effects, particularly in densely built urban areas where buildings are often in close proximity. Zhang et al. [10] highlighted that the aerodynamic performance of chamfered corners can vary significantly under interference conditions, making it crucial to study these scenarios. Understanding how chamfer rates perform in such contexts is vital for optimizing building designs to withstand complex wind interactions in urban environments. Hence, in this study, we perform LBM-LES CFD simulations to investigate how various chamfer rates perform on a building under the interference effect with different wind incident angles.

2 Virtual Wind Tunnel Setup

The setup in this study is based on the wind tunnel experiments conducted at the Tokyo Polytechnic University (TPU), described by Kim et al. [11]. The study focuses on the effect of two identical square buildings without corner modification with a width, B, of 70 mm and a height, H, of 280 mm. The approach flow's mean velocity, U_{ref}, is 8.2 m/s at the reference height at the top of the principal building, and the turbulence intensity measured at the same reference point is 20%. This represents an urban wind exposure with a power law exponent of 0.27. A virtual atmospheric boundary layer (ABL) wind tunnel that consists of spires and roughness blocks was constructed based on the guidelines and recommendations from Irwin [12] and Song [13], among others, to replicate this flow characteristic. These physical features within the simulation domain were responsible for generating the inflow turbulence, thereby avoiding the use of synthetic turbulence injection techniques. Figure 1 shows the computational domain and the flow characteristics at the reference location, which are successfully replicated by the simulation.

The computational mesh employed a Cartesian base grid, enhanced by an octree-based refinement technique that progressively refined the mesh closer to solid surfaces. The mesh size was halved between each refinement level. Six levels of mesh refinement (RL1 to RL6) were applied to the simulation domain with the building models, which resulted in the smallest voxel size of 0.547 mm. This corresponds to 128 voxels along the width of the principal building and 512 voxels throughout its height. Figure 2 illustrates the mesh grid.

A uniform velocity of 10 m/s is applied to the inlet. The domain's top and lateral boundaries were treated as slip walls, while the floor was treated as a no-slip wall, consistent with standard ABL simulations as practiced by Wijesooriya [14]. The no-slip condition enforces zero velocity at the wall, allowing shear stresses to form in the fluid, which is crucial for accurately modeling flow behavior. Furthermore, assuming all solid surfaces, including building facades and the ground, to be smooth,

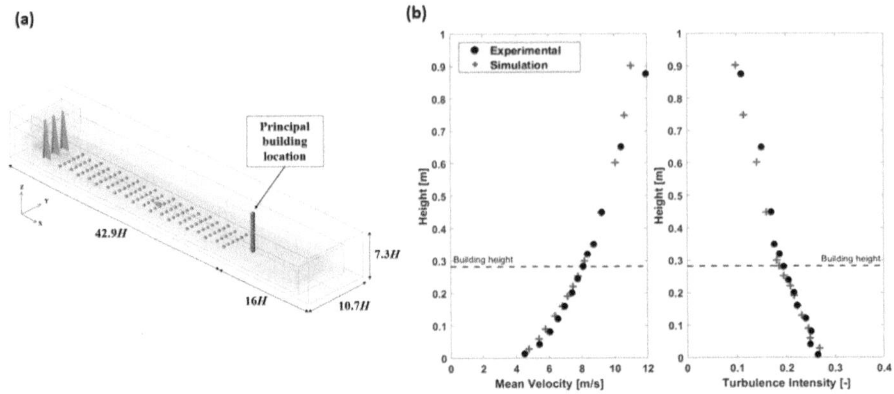

Fig. 1 (**a**) Computational domain of the simulation and (**b**) Flow characteristics at reference location—CFD (present study) vs TPU wind tunnel experiment by Kim et al. [11]

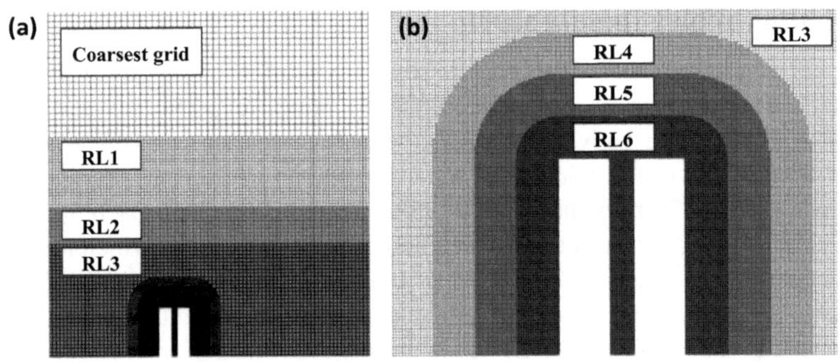

Fig. 2 (**a**) Side view of the mesh grid in the computational domain with (**b**) showing a closer view of the grid near the buildings

the near-wall region was modeled using a Generalized Law of the Wall (GLW) approach with adaptive two-way coupling, without additional roughness effects. Large Eddy Simulation (LES) was conducted using a GPU-accelerated Lattice Boltzmann Method (LBM) based solver. The solver utilizes a cumulant-based collision operator, which improves numerical stability and accuracy in simulating turbulent flows. A simulation time of 12.5 s was set for the solver, where the results of the last 7.5 s were used for the statistical calculations. The simulation was advanced in time using a fixed time step of approximately 2.36×10^{-4} s at the coarsest grid level, resulting in a total of 53,043 time steps for the full 12.5 s simulation. Each simulation case required approximately 24 h of computational time when conducted on eight NVIDIA A100 GPUs.

Chamfers were uniformly applied to all corners of the principal building, while the interference building shape was kept square. The space between the buildings is fixed at 1.5B. The chamfer rate, γ, is defined as the ratio of the chamfer length, b, to the width of the building, B. Six chamfer rates were studied- 0%, 5%, 10%,

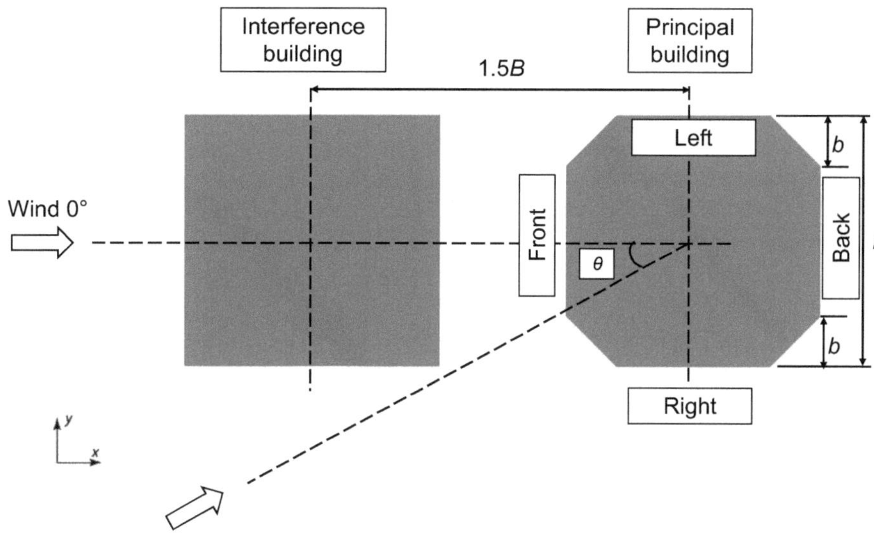

Fig. 3 The plan view of the interference and principal buildings, as well as the parameters

15%, 20% and 25%. Each of these was subjected to five different incident wind angles, θ (0°, 45°, 90°, 135°, and 180°), that create upstream and downstream interference effects. For CFD model evaluation, the pressure coefficient, C_p, from the interference case with a 0% chamfer rate (square model) was compared with experimental data from the TPU Aerodynamic Database website [15]. A flow simulation over an isolated building was also conducted as the reference values to quantify the interference effect. Figure 3 describes the parameters of the investigated interference cases.

Figure 4 presents a scatter plot comparing the mean and RMS pressure coefficients between the TPU experimental data and the current CFD simulation results, using the 45° incident wind case as an example. The overall results, summarized in Table 1, show that a high percentage of data points fall within the 30% tolerance ranges, indicating good agreement between experiment and simulation. A closer examination of height-dependent patterns as shown in Fig. 5 reveals that while $C_{\text{p-mean}}$ errors are uniformly distributed (7–8% of sensors exceeding 10% error at all heights), $C_{\text{p-RMS}}$ errors show greater concentration at mid-height (8%) compared to the top and bottom levels (3–4%). The uniform $C_{\text{p-mean}}$ discrepancies suggest a consistent systematic offset, whereas the mid-height $C_{\text{p-RMS}}$ variations may indicate localized challenges in capturing turbulent flow features. These consistent agreements imply the reliability of the simulation results in this study.

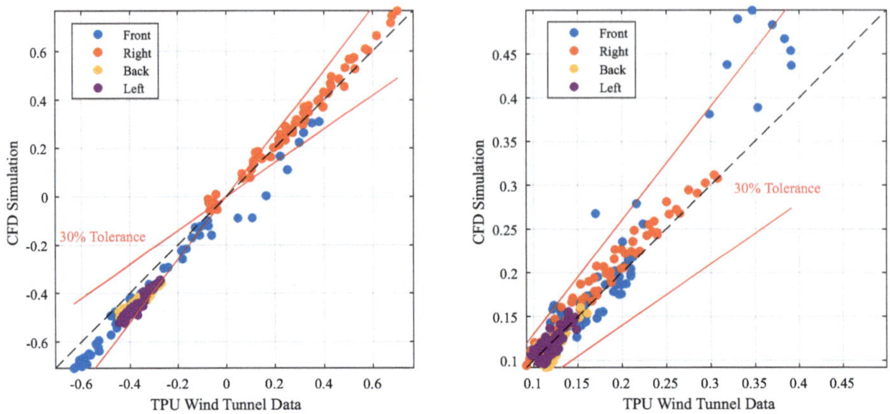

Fig. 4 Scatter plot of mean pressure coefficients, $C_{p\text{-mean}}$, and RMS pressure coefficients $C_{p\text{-RMS}}$

3 Results and Discussion

The effectiveness of the various chamfer rates is discussed in terms of the mean and RMS (fluctuating) base moment coefficients in the along wind, across wind, and torsional components. To quantify the interference effect, the aerodynamic interference factor, AIF, is calculated based on the Eq. 1:

$$\mathrm{AIF} = \frac{\text{base moment coefficient of the building under interference}}{\text{base moment coefficient of the isolated square building}} \tag{1}$$

In principle, base moment coefficients summarize the building's response at its base. They are important for determining the first-order response and equivalent static wind load of the structure. In this study, the results of the mean and the RMS base moment coefficients are determined by:

$$C_{MD} = \frac{\overline{M_D}}{qBH^2}; C'_{MD} = \frac{M'_D}{qBH^2}; \ C'_{ML} = \frac{M'_L}{qBH^2}; C'_{MT} = \frac{M'_T}{qHB^2}; q = 0.5\rho U_{ref}^2 \tag{2}$$

where C_M, C'_M, $\overline{M}$, M', and ρ are the mean base moment coefficient, RMS base moment coefficient, mean base moment, RMS base moment, and fluid density, respectively. The subscripts D, L, and T denote the along wind, across wind, and torsional components, respectively. The results of the interference factor for mean across wind and torsional base moment coefficients are not discussed since these coefficients in isolated cases are close to zero.

Table 1 Percentage of data points within the tolerance for all incident wind angles of interference cases

	0°		45°		90°		135°		180°	
Performance metrics	$C_{p\text{-mean}}$	$C_{p\text{-RMS}}$	$C_{p\text{-mean}}$	$C_{p\text{-RMS}}$	$C_{p\text{-mean}}$	$C_{p\text{-RMS}}$	$C_{p\text{-mean}}$	$C_{p\text{-RMS}}$	$C_{p\text{-mean}}$	$C_{p\text{-RMS}}$
10% tolerance	70.2%	57.9%	18.3%	67.1%	77.8%	66.7%	71.4%	52.8%	41.7%	61.5%
20% tolerance	87.7%	82.9%	60.8%	93.3%	97.2%	96.5%	84.5%	95.7%	83.4%	96.8%
30% tolerance	96.0%	95.6%	90.2%	97.3%	98.9%	99.6%	90.1%	99.7%	97.7%	100%

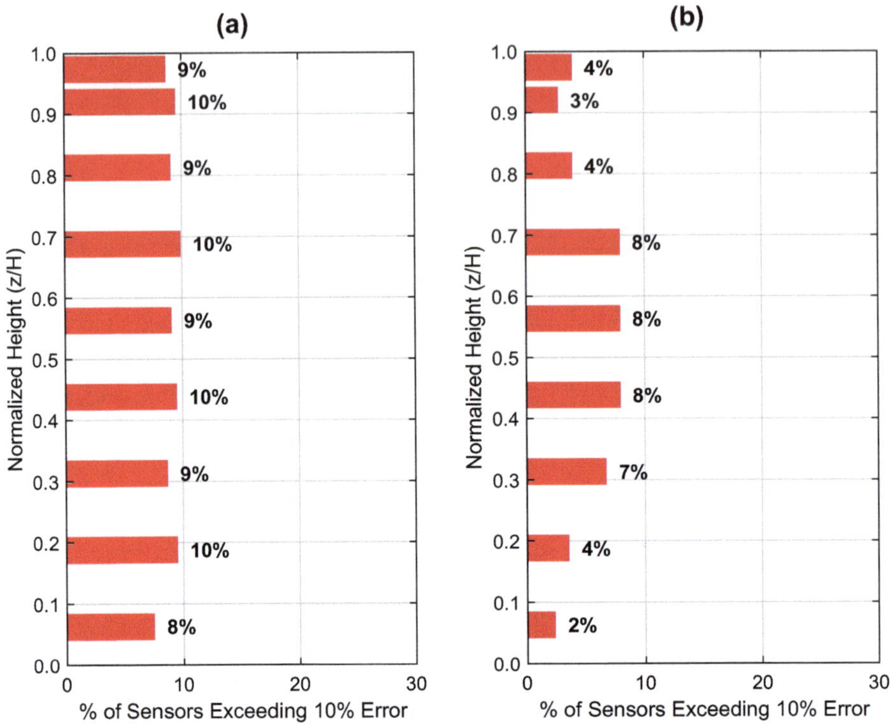

Fig. 5 Percentage of sensors exceeding 10% error by elevation for (**a**) $C_{p\text{-mean}}$ and (**b**) $C_{p\text{-RMS}}$

3.1 Effect of Chamfer Rate in the Along-Wind Direction

Figures 6a, b show the effect of different chamfer rates on the AIF for different incident wind angles in the along wind direction. The AIF of both C_{MD} and C'_{MD} at 0° and 45° wind angles remain consistent across different chamfer rates. This is attributed to the high shielding effect from the interfering building, which significantly reduces the exposure of the principal building to direct wind forces. Consequently, this implies that the effectiveness of chamfering in reducing the forces is limited in the along wind direction when the shielding effect is dominant. According to Kim et al. [11], the shielding effect is pronounced in wind directions ranging from 0° to 75° when the gap between the buildings is very small.

For other incident wind angles, the C_{MD} and C'_{MD} AIF generally decrease as the chamfer rate increases. When the wind flows at 90° and 180°, the minimum C_{MD} AIF is achieved with a 15% chamfer rate, which records approximately a 45% reduction compared to the unmodified square building (0% rate). A similar trend is observed for C'_{MD} AIF at a 90° wind angle, whereas at a 180° wind angle, no significant reduction is observed beyond the 15% chamfer rate.

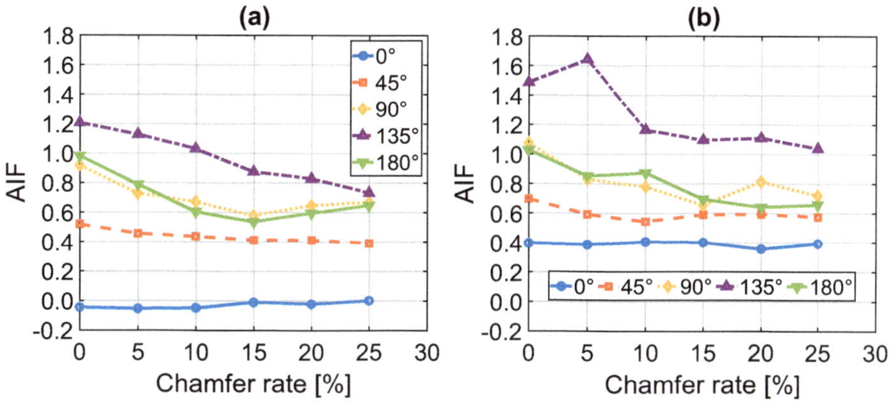

Fig. 6 The AIF of (**a**) C_{MD} and (**b**) C'_{MD} against chamfer rates for different wind angles

It is also evident that the C_{MD} and C'_{MD} values are higher than those in the isolated building case when the wind approaches at a 135° angle compared to other wind directions, as indicated by AIF values greater than 1.0. This observation aligns with the findings by Kim et al. [11], who noted an amplification of the moment coefficient of the principal square building at wind incidence angles between 120° and 165°. This increase is attributed to the channeling effect observed in the interference case and can be effectively mitigated through chamfering. Results show that applying a chamfer rate of 15% successfully reduces the C_{MD} AIF to 0.88, representing a 12% reduction compared to the isolated case. While 5% and 10% chamfer rates do reduce the AIF, they are insufficient as the AIF values still demonstrate an amplification effect relative to the isolated case. The AIF values drop below 1.0 starting from the 15% chamfer rate and decrease further at higher rates. Increasing the chamfer rate allows more freestream fluid to flow into the gap between the buildings. It promotes earlier flow reattachment on the surfaces between the gap, reduces the overall wake downstream, and thus reduces the loads in the along wind direction. This behavior is illustrated by the mean velocity streamlines in Fig. 7.

In contrast, the outcomes are less favorable for C'_{MD}, which reflects the fluctuating component of the force. Despite applying chamfer rates up to 25%, these adjustments are still insufficient to reduce the C'_{MD} below that of the isolated building case. This suggests that the channeling effect has a more significant influence on the dynamic of the forces compared to the mean component, which can be more efficiently mitigated by increasing the chamfer rate. In this study, a reduction to C'_{MD} AIF between 1.04 and 1.17 is maintained when chamfer rates of 10% or higher are applied, while a smaller 5% chamfer rate proves inadequate for this reduction. This may be due to the smaller dimension of the 5% rate being insufficient to facilitate the flow reattachment that separated at the leading edge.

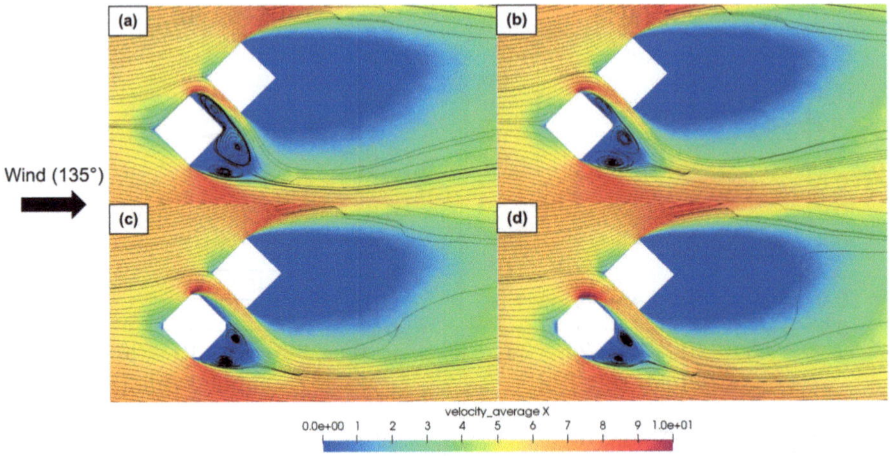

Fig. 7 Mean velocity streamlines for the case of 135° incident wind for chamfer rates, γ, (**a**) 0%, (**b**) 5%, (**c**) 10%, and (**d**) 20%

3.2 Effect of Chamfer Rate in the Across-Wind Direction

In the across wind direction, chamfering has noticeable effects on the C'_{ML} AIF across all wind angles, as indicated in Figure 8a. Unlike the along wind direction, where the shielding effect of the interfering building at 0° or 45° is pronounced, this effect is minimal in the across wind direction. This difference arises because the flow, separated at the trailing edge of the interfering building, dynamically re-attaches to the side faces of the downstream principal building, leading to a higher C'_{ML} than C'_{MD}. At both wind directions, the C'_{ML} AIF can be effectively reduced by increasing chamfer rates. This trend aligns with the results presented by Li et al. [16], who noted that the lowest C'_{ML} was achieved with the largest chamfer applied, which was 20% in their study. In this study, the lowest C'_{ML} is recorded at the highest chamfer rate tested, which is 25%.

A similar trend to that discussed previously with C_{MD} AIF is also observed here for the wind incidence angles of 90° and 180°. The smallest C'_{ML} AIF is achieved with a 15% chamfer rate at a 180° incident wind angle, while at a 90° wind angle, no significant reduction is observed beyond this chamfer rate.

A contrasting effect is observed for different chamfer rates at the 135° wind angle. The C'_{ML} AIF starts to increase from a 5% chamfer rate, peaking at 20%, before decreasing again at a 25% chamfer rate. The maximum C'_{ML} AIF is 1.07, which is a slight amplification compared to the isolated case. As the chamfer size increases, the width of the high $C_{p\text{-rms}}$ region on the chamfer surface between the front and right face also increases, as illustrated in Fig. 9.

The increment of $C_{p\text{-rms}}$ is more prominent on the front face, most probably enhanced by multiple flow separation and reattachment points, as shown by the

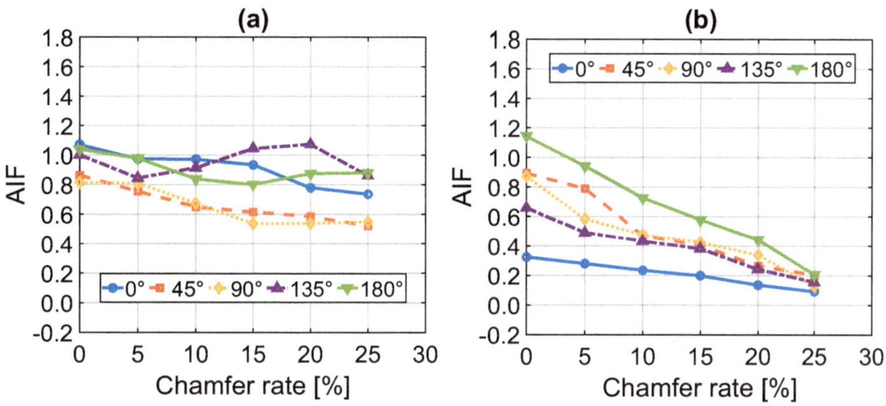

Fig. 8 The AIF of (**a**) C'_{ML} and (**b**) C'_{MT} against chamfer rates for different wind angles

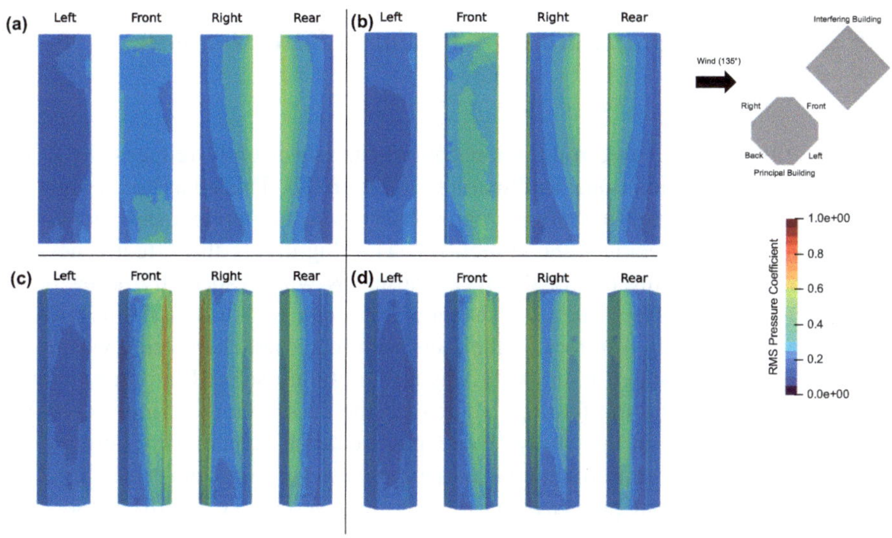

Fig. 9 RMS pressure coefficient at 135° wind for chamfer rates, γ, (**a**) 0%, (**b**) 5%, (**c**) 20%, and (**d**) 25%

mean velocity streamlines in Fig. 10. This is caused by the channeling effect created between the direct surfaces of the two buildings, where more fluids flow into the gap as the chamfer rate increases. Consequently, all of these contribute to the complex flow interactions and add to the higher dynamics generated in the across wind direction.

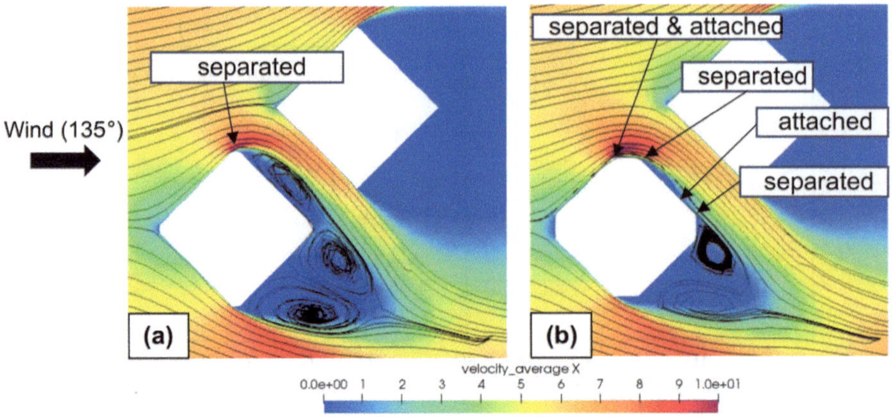

Fig. 10 Mean velocity streamlines plot with the indication of flow separation and reattachment for chamfer rate (**a**) 5% and (**b**) 20%

3.3 Effect of Chamfer Rate on Torsional Base Moment Coefficients

Increasing the chamfer rate for torsional moments consistently reduces the C'_{MT} AIF across all wind directions, as illustrated in Figure 8b. This consistent reduction contrasts with the more fluctuating trends observed in the along wind and across wind directions at specific wind directions. This highlights the effectiveness of chamfering in mitigating torsional effects caused by wind, especially under interference conditions, where such consistency is crucial for structural integrity. A similar trend of reduction in torsional effects with increased chamfering was noted by Li et al. [16]. Although their study was limited to an isolated building under a 0° incident wind angle, the findings align with the observations in more complex interference scenarios in this study. The edges created by chamfering help reduce the torsional forces by minimizing the area exposed to varying wind pressures. This helps to distribute wind pressures more uniformly across the building's surface, thus diminishing the forces that induce twisting motions. Such modifications also enhance the building's ability to withstand torsional stresses, improving overall structural stability and occupant comfort.

4 Conclusions and Future Work

The study evaluated the impact of various chamfered corner modification rates on a building subjected to interference effects under different incident wind angles using LBM-based LES CFD simulations. Chamfering was proven effective in reducing the Aerodynamic Interference Factor (AIF) across all cases, although its effectiveness varied depending on the chamfer rate and wind direction for each loading component. Notably, chamfering significantly reduced the dynamic (RMS) torsional moments across all wind directions but was less effective when the shielding effect from the nearby buildings was dominant in the along wind direction. It is determined that a 15% chamfer rate generally offers the best performance, especially on $C_{MD,}$ and would be a reasonable choice to apply when addressing dynamic forces in along, across, and torsional components. However, an exception was observed for C'_{ML} at 135° wind angle, where 15% and 20% chamfer rates produced slightly adverse effects, highlighting areas for further investigation. These findings provide insights into integrating chamfered corner modifications into architectural designs to enhance aerodynamic performance and structural stability, especially in urban environments where interference effects are pronounced. The CFD model provides a foundation for future research into other corner modifications in similar contexts. Continued investigations are essential to optimize building geometries in response to varied wind conditions, aiming to enhance building safety and occupant comfort in high-rise urban architectures.

Acknowledgments The authors gratefully acknowledge the Gauss Centre for Supercomputing e.V. (www.gauss-centre.eu) for funding this project by providing computing time on the GCS Supercomputer HAWK at Höchstleistungsrechenzentrum Stuttgart (www.hlrs.de). We thank Altair Engineering GmbH for providing the LBM-CFD solver license. The first author is also thankful for the scholarship from the Ministry of Higher Education of Malaysia and the Universiti Teknikal Malaysia Melaka (UTeM), Malaysia.

References

1. Kawai, H.: Effect of corner modifications on aeroelastic instabilities of tall buildings. J. Wind Eng. Ind. Aerodyn. **74–76**, 719–729 (1998)
2. Tamura, T., Miyagi, T.: The effect of turbulence on aerodynamic forces on a square cylinder with various corner shapes. J. Wind Eng. Ind. Aerodyn. **83**(1–3), 135–145 (1999)
3. Li, Y., Tian, X., Tee, K.F., Li, Q.-S., Li, Y.-G.: Aerodynamic treatments for reduction of wind loads on high-rise buildings. J. Wind Eng. Ind. Aerodyn. **172**, 107–115 (2018)
4. Yan, B., Li, Q.-S.: Wind tunnel study of interference effects between twin super-tall buildings with aerodynamic modifications. J. Wind Eng. Ind. Aerodyn. **156**, 129–145 (2016)
5. Shang, J., Zhou, Q., Alam, M.M., Liao, H., Cao, S.: Numerical studies of the flow structure and aerodynamic forces on two tandem square cylinders with different chamfered-corner ratios. Phys. Fluids. **31**(7), 75102 (2019)
6. Cao, W., Wang, X., Liu, Y., Yin, Y., Yang, J., An, J.: Large eddy simulation on wind-induced interference effects of staggered chamfered square cylinders. Sci. Rep. **13**(1), 17532 (2023)
7. Ding, F., Kareem, A.: A multi-fidelity shape optimization via surrogate modeling for civil structures. J. Wind Eng. Ind. Aerodyn. **178**, 49–56 (2018)
8. Elshaer, A., Bitsuamlak, G., El Damatty, A.: Enhancing wind performance of tall buildings using corner aerodynamic optimization. Eng. Struct. **136**, 133–148 (2017)
9. Adeeb, E., Haider, B.A., Sohn, C.H.: Flow interference of two side-by-side square cylinders using IB-LBM—effect of corner radius. Results Phys. **10**, 256–263 (2018)
10. Zhang, J., Li, F., Zhang, Z., Zhang, T., Wang, C., Xiang, B., Zhang, Y.: Wind-induced interference effect of Chamfered Square cylinders in tandem and side-by-side arrangements. Buildings. **12**, 2125 (2022)

11. Kim, W., Tamura, Y., Yoshida, A.: Interference effects on aerodynamic wind forces between two buildings. J. Wind Eng. Ind. Aerodyn. **147**, 186–201 (2015)
12. Irwin, H.P.A.H.: The design of spires for wind simulation. J. Wind Eng. Ind. Aerodyn. **7**(3), 361–366 (1981)
13. Song, P.: Scholarship at UWindsor simulation of atmospheric boundary layer in an open-loop wind tunnel using spire-roughness-element technique simulation of atmospheric boundary layer in an open-loop wind tunnel using spire-roughness-element technique by. (2017)
14. Wijesooriya, K., Mohotti, D., Chauhan, K., Dias-da-Costa, D.: Numerical investigation of scale resolved turbulence models (LES, ELES and DDES) in the assessment of wind effects on supertall structures. J. Build. Eng. **25**, 100842 (2019)
15. TPU (Tokyo Polytechnic University) Aerodynamic Database.: Two adjacent high-rise buildings with square plan (2011). https://db.wind.arch.t-kougei.ac.jp/
16. Li, Y., Li, C., Li, Q.S., Song, Q., Huang, X., Li, Y.G.: Aerodynamic performance of CAARC standard tall building model by various corner chamfers. J. Wind Eng. Ind. Aerodyn. **202**, 104197 (2020)

Nonlinear Performance of a 20-Story Steel Building Under Recurring Seismic and Wind Loads

Shairin Islam, Asad Ullah, and Anastasia Athanasiou

Abstract This paper provides insight in the response of a 20-storey steel structure subjected to recurring seismic and wind loads, through collapse. The seismic and wind assessment performance framework applied herein, is based on the comprehensive procedures recommended by FEMA P695 and the ASCE Prestandard for Performance-Based Wind Design. The far-field FEMA set of strong ground motion excitation is used for seismic simulations. Alongwind and crosswind loads, derived from local aerodynamic data available at the Tokyo Polytechnic aerodynamic database, are used in wind simulations. Detailed, lumped plasticity models of the lateral force resisting system are developed in OpenSees. The numerical models are used to perform nonlinear simulations that predict with sufficient accuracy the multi-hazard structural performance, through collapse. Seismic and wind Incremental Dynamic Analyses provide insight into the near collapse behaviour of the building and allow quantification of the margin of safety against collapse under extreme events. System performance is quantified in terms of peak drift, floor acceleration, and residual drift. The building, fails to meet drift and acceleration criteria under service level winds. The near collapse structural response under very strong motion excitation is characterised by significant damage, distributed over multiple stories; whereas damage under ultimate wind loads is limited at the lower floors, where wind demands are dominant. Yielding under alongwind and crosswind loads occurs for events characterised by high (compared to design) wind speeds. The results reinforce the belief that wind design would benefit greatly from the introduction of limited inelasticity and inherent overstrength.

Keywords Multi-hazard assessment · Nonlinear analysis · Incremental dynamic analysis · Seismic loads · Wind loads

---

S. Islam · A. Ullah · A. Athanasiou (✉)
Bauhaus-University Weimar, Weimar, Germany
e-mail: anastasia.athanasiou@uni-weimar.de

I. Calotescu, C. I. Coşoiu (eds.), *Proceedings of the Third National Conference on Wind Engineering*, Springer Proceedings in Earth and Environmental Sciences,
https://doi.org/10.1007/978-3-032-09671-5_4

1 Introduction

Natural hazards, such as strong earthquakes and extreme windstorms are responsible for significant annual losses worldwide. Performance-based methodologies, were first developed for seismic applications, allowing quantification of the system performance under extreme events [1]. They also provide insight into the collapse limit state response, a state associated with both human casualties and economic losses (structural and non-structural damage, repair times and costs). Current seismic design procedures, following major international building codes, have used the knowledge gained from extensive research, experimentation, and observation of past events, to provide a structurally efficient and economic seismic design. The fundamental concepts of modern seismic design are ductility and capacity design. Structures are designed to undergo controlled inelasticity (ductility) at predetermined well-detailed members (capacity design) under the design earthquake, occurring once in the mean recurrence interval (MRI). The premise of Performance-Based Seismic Design (PBSD) is to provide buildings that meet specified performance goals at specified limit states, rather than strictly adhering to prescriptive design details.

Disastrous wind events in the 1990s motivated the development of performance-based methodologies for the assessment and design of wind-excited structures [2]. However, due to significant differences in the nature of winds and earthquakes, and their effects on structures, performance-based wind engineering has developed less rapidly than performance-based seismic engineering. Wind events can last for hours, and require prolonged resistance from structures, while earthquake loads are intense but short. Under wind, the structure undergoes numerous inelastic deformations over time, which may trigger failure through low-cycle fatigue, whereas earthquakes induce less inelastic excursions to the system. Wind design following modern building codes is prescriptive and has provided the society with buildings of unquantifiable reliability. Moreover, the requirement of elasticity under the design wind loads (Ultimate Limit State - ULS) and the satisfaction of stringent performance criteria under frequent winds (Serviceability Limit State - SLS) does not promote the efficient and economical design of taller buildings. In this light, in 2019 ASCE published the Prestandard for Performance-Based Wind Design (PBWD), that together with its recent update in 2023, provides a viable design alternative to the current, prescriptive wind design [3].

Recent performance-based design methods highlight the importance of integrating multiple hazards like wind and earthquake by: (*i*) using wind tunnel data for demand evaluation, (*ii*) employing ductility concepts for limited-ductility LFRS design considering wind loads [4], (*iii*) implementing probabilistic and deterministic wind assessments [5–9], (*iv*) mapping functionality to LFRS and building envelope limit states under combined hazards [6, 8, 9], and (*v*) using loss estimation models for multihazard decision-making [6].

This study focuses on the assessment of the performance of a 20-story steel building in urban California, under recurring strong motion, alongwind and crosswind excitation. The findings presented herein, contribute to the current state of knowledge in PBWD by: (*i*) using wind tunnel data for reliable representations of alongwind and crosswind loads acting on buildings; (*ii*) performing nonlinear simulations under extreme seismic and wind events; (*iii*) assessing the margin of safety against collapse of buildings sited in combined seismic and wind environments.

2 Methodology

This paper presents a comprehensive approach to assess the performance of tall steel buildings under dynamic conditions, based on the FEMA P695 [10] and the ASCE Prestandard for PBWD [3]. Since the earthquakes and winds have very low probability of concurrence, they are considered as independent events and the corresponding analyses are performed in parallel. Figure 1 illustrates the seismic assessment framework considered herein. The assessment begins with the hazard characterization and continues with the collection of strong ground motion data. The next step is to develop a detailed finite element model, that accounts for material and geometrical nonlinearities. Response history analyses at the design level and beyond provide insight into the seismic behavior, through collapse. Incremental Dynamic Analysis (IDA) shows how the system responds to seismic excitation of increasing intensity [11]. The spectral acceleration at the fundamental period $S_a(T_1)$ is commonly used to characterize the seismic intensity. Peak and residual interstory drifts, and peak floor accelerations are often considered as the Engineering Demand Parameters (EDP) of interest. The IDA curves show the evolution of the EDP as a function of the seismic intensity, $S_a(T_1, 5\%)$. They are also used to determine the Collapse Margin Ratio, CMR, defined as the ratio of the seismic intensity that causes collapse, S_{CT}, to the design spectral acceleration at the first period T_1, S_d $(T_1,5\%)$: CMR $= S_{CT}$ $(T_1,5\%)/$ S_d $(T_1,5\%)$. CMR is a key metric defined in FEMA P695, that evaluates how much stronger ground motions must be compared to design level earthquakes to cause collapse.

Figure 2 schematically outlines the wind performance assessment process, mirroring the earthquake assessment procedure detailed in Fig. 1. The assessment begins with the wind hazard analysis and the wind tunnel-derived loads. A detailed nonlinear finite element model is then developed. Response history analyses, up to collapse, reveal the wind behavior. Incremental Dynamic Wind Analysis (IDWA) is the result of the growing interest in PBWD. Three challenging aspects of performance-based wind assessment frameworks are: (*i*) deriving reliable wind

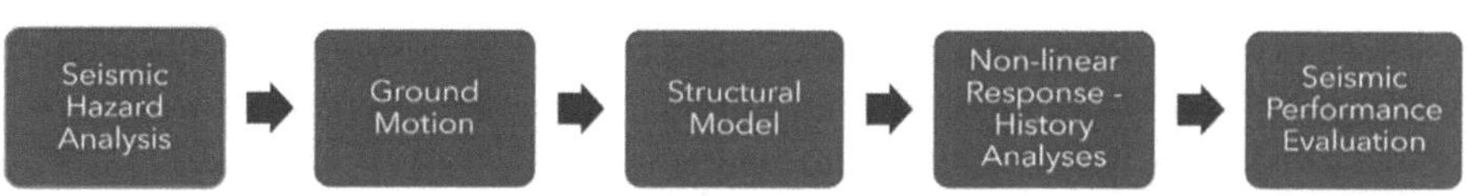

Fig. 1 Brief overview of the applied performance-based, seismic assessment framework

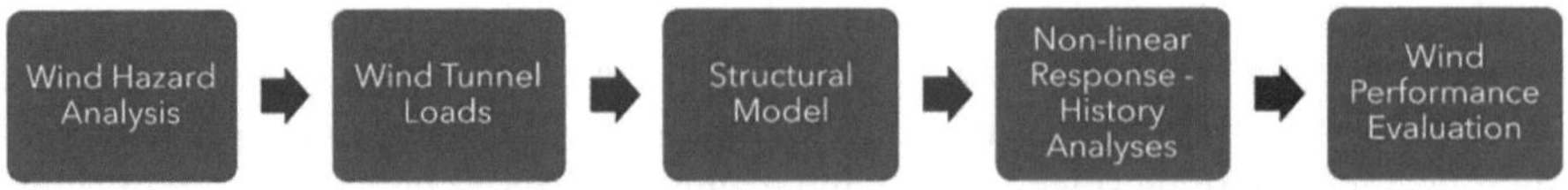

Fig. 2 Brief overview of the applied performance-based, wind assessment framework

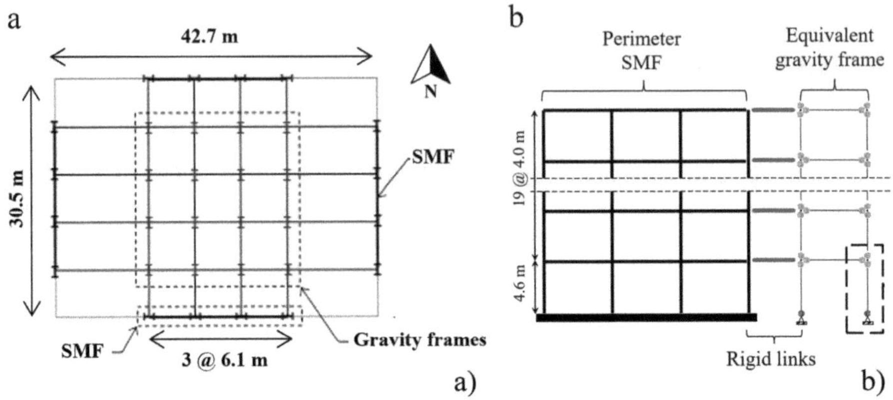

Fig. 3 (**a**) Typical floor plan, and (**b**) 2D numerical model for ½ of the floor plan (adapted from [17])

loads, (*ii*) conducting nonlinear wind analysis, and (*iii*) scaling winds for IDWA. In IDWA, wind loads are linearly scaled on the basis of the wind intensity, typically the mean wind speed at 10 m, and input to excite the structural model. Whereas linear scaling of wind excitation on the basis of the mean speed is discouraged, this approach has become a common practice in IDWA applications [4–6, 12, 13].

3 Case Study

3.1 Building Description

The case study is a 20-story frame steel building, Risk Category II, sited in a combined seismic and wind environment in urban California (Seismic Design Category: $S_{DC}=D_{max}$, soil class D). This is an archetype building studied extensively in NIST [14], Elkady and Lignos [15, 16]. The building is subjected to California's seismic demands in a generic urban California setting. The key seismic and wind hazard parameters are based on typical conditions for places like downtown San Francisco, but no specific street, or city is defined. Figure 3a, b show the typical floor plan and building elevation. The building is regular in plan and elevation. The first and typical story heights are 4.6 m and 4 m, respectively.

Table 1 Seismic design parameters for the 20-story SMF archetype building design

S_{DC}	R	Ω_o	T	V/W	S_{MT}	S_{M1}	S_{D1}
D_{max}	8	3	3.37 s	0.037	0.27 g	0.90 g	0.60 g

The three-bay Special Moment Frames (SMFs) in the building perimeter are designed to withstand the lateral loads in the N-S and E-W directions, respectively. A SMF in the context of ASCE/SEI 7–10 [18] is a moment frame designed to resist lateral seismic forces through flexural strength of beams and columns and rigid beam-column connections. SMFs sustain large inelastic deformations without collapse during severe seismic events. They are detailed as per ASCE/SEI 7–10 [18] and AISC 341–10 [19]: beam-column connections are designed to accommodate plastic hinges (in beams) without premature failure, panel zones are strengthened to prevent shear failure in beam-column joints; and column splices are located away from high-stress regions and designed for full moment capacity.

Table 1 summarizes the key seismic design parameters for the SMF building. The values of the seismic base shear coefficient, *V/W*, and the Maximum Considered Earthquake (MCE) spectral acceleration, S_{MT}, the MCE and design earthquake (D) 1-second spectral accelerations, S_{M1} and S_{D1} for SDC D_{max} were determined using the ASCE/SEI 7-based structural period, *T*, and the spectral values given in Table 5–1 of FEMA P695. ASCE/SEI 7 assigns SMFs a response modification coefficient $R = 8$, which accounts for high ductility and energy dissipation capacity, and an overstrength factor $\Omega_o = 3$ for the design of force-controlled members.

Here, beams and columns are made of steel ASTM A992 Gr. 50 ($F_{y,nominal} = 345$ MPa). The expected steel yield stress (i.e., 380 MPa) is used for the seismic design of beams, whereas the minimal specified yield stress (i.e., 345 MPa) is used for the seismic design of columns and panel zones. The columns are spliced at mid-height of odd-numbered stories, and fixed at the foundation level. Beam-to-column connections are fully-restrained with Reduced Beam Sections (RBS), as per AISC 358–10 [20]. The panel zones resist significant shear, tension and compression forces from the beams framing to the column. Since panel zones are characterized by high ductility, design shall accommodate balanced yielding between the plastic hinge zones in beams and the panel zones. Details of the building design along the pushover curves of the building can be found in Elkady and Lignos [15, 16]. Elkady [17] notes that wind governs the response of the 20-story building, but does not provide details of the wind design.

3.2 Structural Model

Due to the symmetry of the building, two numerical structural models are developed in the two orthogonal directions for 1/2 of the building plan. Each 2D model is developed in OpenSees and includes a perimeter SMF in the direction of interest,

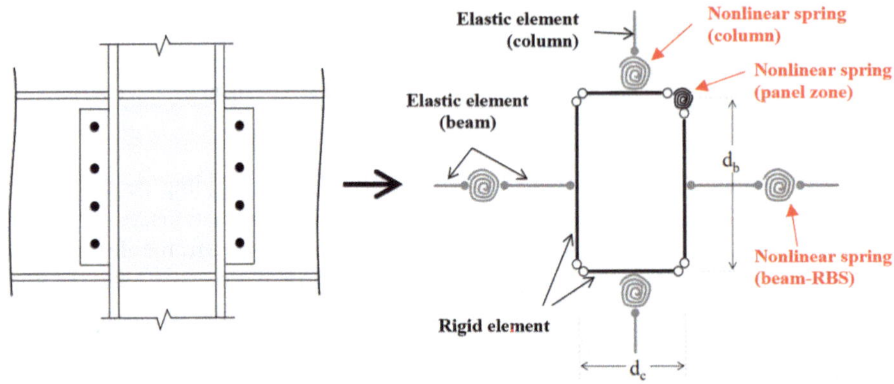

Fig. 4 Plasticity model of Beam-to-Column Panel Joint (adapted from [17])

as shown in Fig. 3b. The models account for inherent overstrength, material nonlinearity, and P-delta effects. The modified Ibarra-Medina-Krawinkler (IMK) model, calibrated using the regression equations of Lignos and Krawinkler [21], models the moment and rotation behavior of the bare beam. The panel zone shear force-shear distortion in bare steel beam-to-column connections is modeled using parameters calculated based on PEER/ATC 72–1 [22]. Note that the steel column deterioration model does not consider deterioration due to axial shortening and lateral torsional buckling. The nonlinear behavior of columns, beams, and panel zones is simulated using a concentrated plasticity approach. This method idealizes members as elastic elements with rotational flexural springs at potential plastic hinge locations. Fig. 4 shows the plasticity model of Beam-to-Column Panel Joint, i.e., the parallelogram model proposed in Gupta and Krawinkler [23]. *P-Delta* effects are simulated using the *leaning column*.

The leaning column is connected to the SMF using axially-rigid links, as shown in Fig. 3b. A fixed support is assumed for the beam-column base. Rayleigh damping is assigned to the elastic elements, considering a damping ratio of 2% for the first and third modes, following the recommendations of Zareian and Medina [24]. The Krylov-Newton solution algorithm is used for nonlinear analysis in OpenSees. Eigenvalue analysis provides the first three modal periods: $T_1 = 3.57$ s, $T_2 = 1.23$ s, $T_3 = 0.68$ s.

3.3 Seismic Performance Assessment

Figure 5 illustrates the response spectra for the FEMA P695 far-field motion set, that was developed as part of the ATC-63 project. The set includes 22 pairs (44 individual) horizontal ground motion records selected to represent strong shaking from magnitude 6.5 to 7.9 earthquakes. Intended to represent earthquake ground

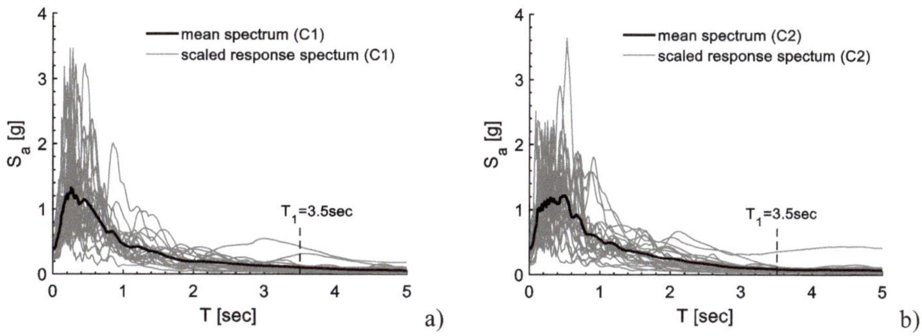

Fig. 5 FEMA P695 far-field motion set scaled response spectra: (**a**) C1, (**b**) C2 components

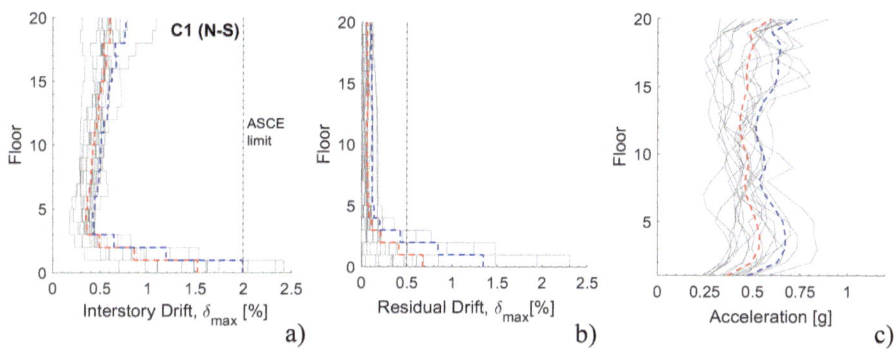

Fig. 6 Response to the scaled FEMA P695 far-field motions (C1, N-S direction) in terms of: (**a**) peak interstory drift, (**b**) residual drift, and (**c**) peak floor acceleration. Red and blue dashed lines indicate μ and $\mu + \sigma$ response, respectively

motions recorded at distances greater than 10 km from the fault rupture, these records are utilized in IDA to assess the collapse safety of structural archetypes under seismic loading. For seismic analyses the two orthogonal components of each ground motion pair, C1 and C2, are applied in the N-S and E-W directions of the building, respectively.

Figures 6 and 7 show the peak response of the system in the N-S and E-W directions, in terms of peak interstory drift, residual drift, and peak floor accelerations, under the scaled 22 FEMA P695 far-field motion pairs (C1, C2). The mean (μ) and mean plus one standard deviation ($\mu + \sigma$) of the peak responses are also shown in the plots. Here, the ground motions were normalized using scaling factors recommended by FEMA, resulting in mean and median peak ground velocities of approximately 80 cm/s, and a 5% damped spectral acceleration at a 1.0-second period, S_a (1 s, 5%), approximately equal to the 1-second design spectral acceleration, $S_{D1} = 0.60$ g. Figures 6a, b show how the interstory and residual drifts are higher in the lower stories of the building. The peak drift is within the ASCE/SEI 7 limit of 2.0% for buildings in Risk Category II; however, the peak residual drift

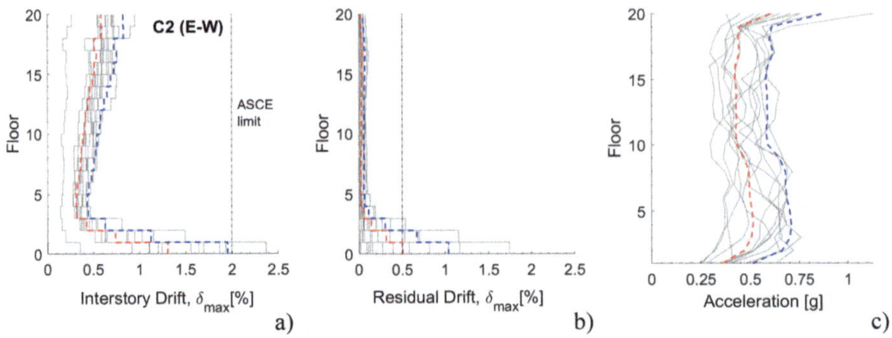

Fig. 7 Response to the scaled FEMA P695 far-field motions (C2, E-W direction) in terms of: (**a**) peak interstory drift, (**b**) residual drift, and (**c**) peak floor acceleration. Red and blue dashed lines indicate μ and $\mu + \sigma$ response, respectively

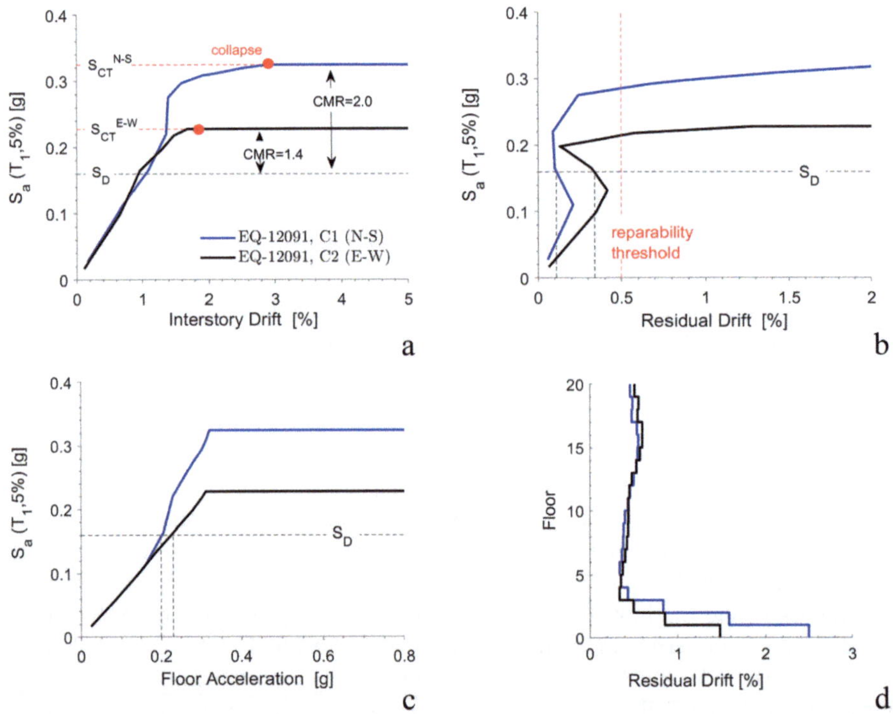

Fig. 8 (**a–c**) Seismic IDAs in the N-S and E-W directions, with the EDP being peak interstory drift, residual drift, peak floor acceleration, and (**d**) residual drifts in the N-S and E-W directions scaled to near collapse

exceeds the reparability threshold of 0.50%. Residual drift is a direct indicator of damage; high values indicate that the building may not be safe to occupy without prior reparation works or evaluation for structural integrity. Figure 6c, which shows the peak floor accelerations, also indicates significant nonlinear behavior induced by seismic excitation. A comparison of Figs. 6 and 7 reveals that the seismic behaviour of the building is similar in the two orthogonal directions.

Figures 8a–c show the IDA curves in the N-S and E-W directions, with the EDP being: peak interstory drift, residual drift, and peak floor acceleration. Due to time constraints, the IDA was performed on only one pair of horizontal motions, the EQ-12091 pair. Figure 8a reveals how the structure experiences substantial inelasticity prior to collapse. The spectral acceleration at collapse, S_{CT}, is 0.32 g under C1 (applied in the N-S direction), and 0.23 g under C2 (applied in the E-W direction). This corresponds to a CMR of 2.0 in the N-S direction, and 1.4 in the E-W direction, considering that the design spectral acceleration is S_D (T_1, 5%) = 0.16 g. This demonstrates that the structure has a significant margin of safety against the seismic demand induced by motion EQ-12091 (C1, C2). Nevertheless, given that only one pair of motions is considered herein, no further conclusions can be drawn. Fig. 8b shows the IDA curves in the N-S and E-W directions, expressed in terms of S_a (T_1,5%) and residual drift. It is again clear that the structure experiences increased permanent deformation under C2, in the E-W direction. Fig. 8c illustrates the IDA curves for the two motion components, expressed in terms of S_a (T_1,5%) and floor acceleration. The two components induce similar acceleration demands up to an intensity S_a close to the design spectral acceleration, S_D. Finally, subplot 8d shows the residual drifts in the N-S and E-W directions of the building, under the C1 and C2 components of the ground motion EQ-12091, scaled to near collapse. A small increase in the seismic intensity, S_a, leads the system to failure. The damage is distributed throughout the lower half of the building, with hinges developing at all beams and at the base of the first floor columns.

3.4 Wind Performance Assessment

A set of 20 alongwind and a set of 20 crosswind loads are obtained from local pressure data available at the TPU aerodynamic database (https://wind.arch.t-kougei.ac.jp/system/eng/contents/code/tpu). These loads are applied along the height of the structural model, at the floor levels, to perform simulations of the response to alongwind and crosswind excitation. Due to the computational time required for wind history analysis, only one wind data set is considered in this study. The TPU aerodynamic data consists of pressure coefficient histories, $C_pC_g(t)$, recorded during wind tunnel tests on the four sides of a building model, that is geometrically similar to the case study (*H-B-D* = 2–1-1, exponent factor a = 1/4). Since the TPU pressure coefficients refer to the scaled model, they have to be scaled in time using the amplification factor VS/LS, where LS = 1:400 and VS = 1:6 are the length and velocity scales. Such a scaling procedure ensures similitude between the Strouhal number at the model and at full scale. The time step in the tests is 0.001 s, whereas at full scale it is dt = 0.2 s. While the wind tunnel tests were performed at various attack angles, the focus here is on the wind direction

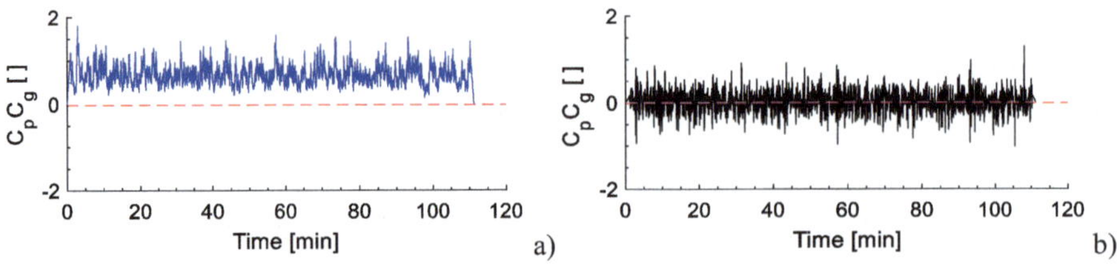

Fig. 9 (**a**, **b**) Alongwind and crosswind pressure coefficients recorded at the mid-height of the reference TPU building model (H-B-D = 2–1-1, exposure factor a = 1/4, wind angle = 0°)

Table 2 Wind design speeds for San Francisco, specified for 3-second and 1-hour durations at a standard height of 10 meters over open, flat terrain. Values are scaled to various MRI

MRI	10	50	700	1700	10,000	100,000	10^6	years
U_{3s}	28	33	41	44	50	57	66	m/s
U_{3600s}	18.4	21.7	27	28.9	32.9	37.5	43.4	m/s

perpendicular to the building facade (0° angle of attack). Fig. 9 shows the alongwind and crosswind C_pC_g coefficients recorded at the mid-height of the tested model. The pressure coefficients are then integrated into the wind force histories acting on the various floor levels:

$$F(t) = C_pC_g(t)\, A\, q \ , \quad q = 1/2\, \rho U^2 \tag{1}$$

where A is the corresponding tributary floor area, q is the reference wind pressure, $\rho = 1.29\ kg/m^3$ is the standard atmospheric pressure, and U is the mean wind speed at the reference height (typically 10 m or the roof height H).

Herein, alongwind and crosswind forces are calculated at all floor levels, and are scaled to the limit state of interest (i.e., SLS and ULS) on the basis of the mean wind speed U. The online platform https://ascehazardtool.org/ provides the 3-s gust wind speeds at 10 m above ground for Exposure C Category (open flat terrain) for any location in US. The wind speeds correspond to a ~ 7% probability of exceedance in 50 years (annual exceedance probability = 0.00143, MRI = 700 years). Table 2 provides the 3-second gust wind speeds, U_{3s}, for San Francisco at various MRIs: 10-, 50-, 700-, 1700-, 10,000-, 100,000, and 10^6-years. These values are divided by a conversion factor of 1.52 to provide the corresponding hourly mean wind speed at 10 m U_{3600s} [25].

Consideration of the above and observation of Fig. 9 leads to the following conclusions: (*i*) the wind duration is approximately 2 hrs, i.e., which is significantly longer than the duration of a ground motion; (*ii*) the alongwind component has a significant mean component; (*iii*) the crosswind component has zero mean, and consists only of the background and resonant components; and finally (*iv*) wind loads depend on both the environment and the building geometry, whereas earthquakes are only typical of the site. A 60-s wind force ramp is added to each wind force history. This allows the wind force amplitude to increase smoothly from

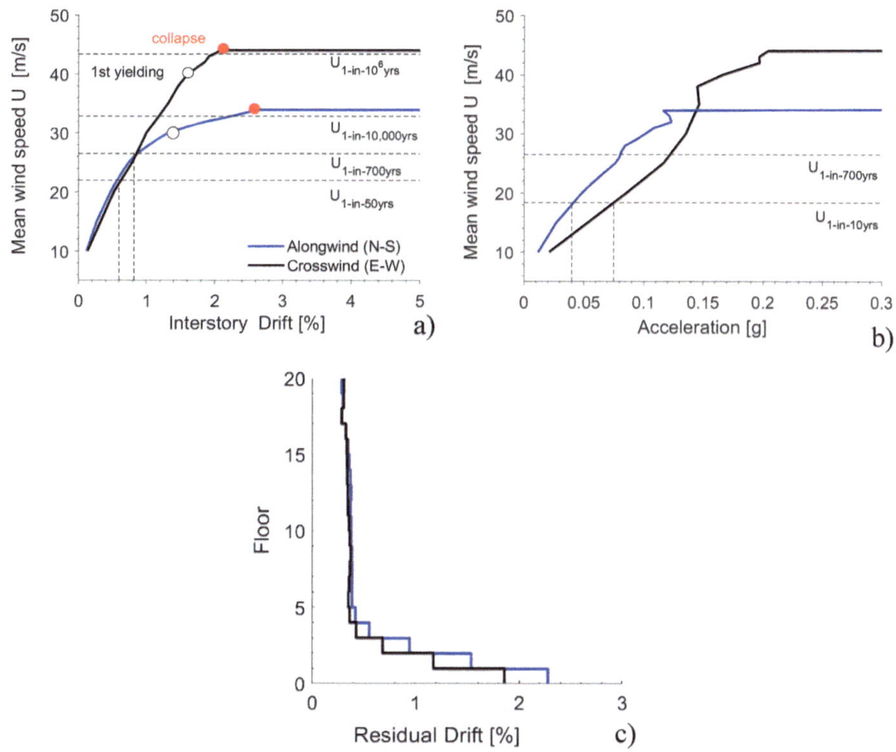

Fig. 10 (**a**, **b**) Wind IDAs in the N-S and E-W directions, with the EDP being peak interstory drift, peak floor acceleration, and (**c**) residual drifts in the N-S and E-W directions, scaled to near collapse

zero to $F_W(t = 0 \text{ s})$. In the case of alongwind excitation, $F_W(t = 0 \text{ s}) > 0$, and the addition of the ramp avoids overestimating the system response due to transient effects.

Figures 10a, b show the IDWA curves under alongwind and crosswind excitation, where the mean wind speed at 10 m height, U_{3600s} or simply U, is expressed as a function of the interstory drift and the floor acceleration, respectively. Subplot 10 a shows how the system behaves linearly even beyond the 1-in-700 years, design wind. The design level (ULS) considers structural safety, and refers to a wind event with $U = 27$ m/s and MRI $= 700$ years. SLS, on the other hand, is concerned with conditions for building usability, and is associated to wind events with MRI $= 10$, 50 years and mean wind speed $U = 18.4$ and 21.7 m/s. Collapse in the alongwind and crosswind directions occurs under very rare winds with MRI > 10,000 years and 10^6 years respectively. Alongwind excitation is more critical for the investigated building. However, the taller the building, the more likely it is that crosswind excitation will become critical.

Both alongwind and crosswind components induce similar drift demands at the SLS and ULS, approximately equal to 0.60% and 0.80% respectively. These drift demands significantly exceed the 1/400 (1-in-50 year SLS wind) and 1/200 (1-in-700 year ULS wind) thresholds suggested by the ASCE Prestandard for PBWD for Risk Category II buildings. The acceleration demand at SLS (1-in-10 years wind) is also excessive, ranging from 0.04 g to 0.075 g for the alongwind and crosswind components, respectively. This means that the building is wind sensitive, and cannot meet the occupant comfort and serviceability criteria prescribed by the ASCE Prestandard for PBWD. Subplot 10 c presents the residual drifts under alongwind and crosswind events scaled to near collapse. Damage is concentrated in the lower floors, where the wind demands are dominant. In fact, near collapse, plastic hinges form at the lower floor beams and at the base of the ground floor columns.

4 Conclusions

This paper presents a performance-based assessment methodology for tall buildings subjected to both ground motion and wind excitation. The framework is illustrated through the assessment of an archetype 20-story steel SMF building in urban California, designed for seismic and wind loads. Detailed finite element models of the LFRS in the two orthogonal directions are developed in OpenSees. The models account for material and geometrical nonlinearities. Nonlinear response simulations are performed by subjecting the LFRS models to the FEMA P695 far field set and alongwind and crosswind loads derived from available local aerodynamic data on a geometrically similar building. Seismic and wind IDA provide insight into the dynamic building behavior, through collapse. Despite the limited scope of the IDA performed (in terms of the number of ground motion records and wind sets considered), the results obtained are consistent with the expected behavior of the building. Under seismic loads, scaled near the collapse limit state, the system experiences significant nonlinearity distributed over the lower half of the building. Under wind loads scaled to near collapse intensity, the system experiences damage only in the lower four floors where the demands are critical. Other important findings are that the SLS wind criteria are not met, both in terms of peak drift and peak floor acceleration. First yielding in the critical alongwind direction requires very rare events (MRI > 700 years), and collapse requires even rarer winds (MRI > 10,000 years). Unlike the more ductile seismic response (which justifies inelastic design codes), the wind response does not justify linear design. The building economy and structural efficiency would benefit from consideration of limited inelasticity and inherent system overstrength under design level winds. While acknowledging the limitations of the current findings, this study reinforces the importance of future nonlinear wind simulations, particularly for taller buildings in diverse locations. Furthermore, it highlights the potential for adapting and validating PBSD analytical concepts and numerical tools for PBWD.

Acknowledgements The authors would like to acknowledge the financial support to the second and third author awarded from the Carl-Zeiss foundation.

References

1. Whittaker, A., Hamburger, R., Mahoney, M.: Performance-based engineering of buildings for extreme events. In: Proceedings of the AISC-SINY symposium on resisting blast and progressive collapse, pp. 55–66. American Institute of Steel Construction, New York, NY (2003)
2. Ellingwood, B.R., Rosowsky, D.V., Li, Y., Kim, J.H.: Fragility assessment of light-frame wood construction subjected to wind and earthquake hazards. J. Struct. Eng. **130**, 1921–1930 (2004)
3. ASCE: Pre-Standard for Performance-Based Wind Design. American Society of Civil Engineers, Reston, Virginia (2019, 2023)
4. Athanasiou, A., Tirca, L., Stathopoulos, T.: Performance-based wind and earthquake design framework for tall steel buildings with ductile detailing. J. Wind Eng. Ind. Aerodyn. **240**, 105492 (2023)
5. Athanasiou, A., Tirca, L., Stathopoulos, T.: Nonlinear wind and earthquake loads on tall steel braced frame buildings. ASCE J. Struct. Eng. **148**(8), 04022098 (2022)
6. Athanasiou, A., Dakour, M., Pejmanfar, S., Tirca, L., Stathopoulos, T.: Multihazard performance-based assessment framework for multi-story steel buildings. ASCE J. Struct. Eng. **148**(6), 04022054 (2022)
7. Kwag, S., Gupta, A., Baugh, J., Kim, H.: Significance of multi-hazard risk in design of buildings under earthquake and wind loads. Eng. Struct. **243**, 112623 (2021)
8. Li, C., Pan, H., Tian, L., Bi, W.: Lifetime multi-hazard fragility analysis of transmission towers under earthquake and wind considering wind-induced fatigue effect. Struct. Saf. **99**, 102266 (2022)
9. Rizzo, F., Caracoglia, L., Maddaloni, G., Sabbà, M.F., Foti, D.: Exploring multi-hazard effects on a tall building and its non-structural elements through simultaneous earthquake and wind loading. J. Build. Eng. **91**, 109489 (2024)
10. FEMA: Quantification of building seismic performance factors. In: FEMA P695. FEMA, Washington, DC (2009)
11. Vamvatsikos, D., Cornell, C.A.: Incremental dynamic analysis. Earthquake Eng. Struct. Dynamics. **31**(3), 491–514 (2002)
12. Athanasiou, A., Tirca, L., Stathopoulos, T.: Scaling wind loads for incremental dynamic analysis applications. J. Wind Eng. Ind. Aerodyn. **263**, 106116 (2025)
13. Venanzi, I., Lavan, O., Ierimonti, L., Fabrizi, S.: Multihazard loss analysis of tall buildings under wind and seismic loads. Struct. Infrastruct. Eng. **14**(10), 1295–1311 (2018)
14. NIST: Evaluation of the FEMA P695 Methodology for Quantification of Building Seismic Performance Factors. NIST GCR 10-917-8. NEHRP consultants Joint Venture (2010)
15. Elkady, A., Lignos, D.G.: Effect of gravity framing on the Overstrength and collapse capacity of steel frame buildings with perimeter special moment frames. Earthquake Eng. Struct. Dynamics. **44**(8), 1289–1307 (2015)
16. Elkady, A., Lignos, D.G.: Modelling of the composite action in fully restrained beam-to-column connections: implications in the seismic design and collapse capacity of steel special moment frames. Earthquake Eng. Struct. Dyn. **43**(13), 1935–1954 (2014)
17. Elkhady, A.: Collapse risk assessment of steel moment resisting frames designed with deep wide-flange columns in seismic regions. PhD dissertation, McGill University, Montreal (2016)
18. ASCE: Minimum Design Loads and Associated Criteria for Buildings and Other Structures. ASCE 7-10. American Society of Civil Engineers, Reston, VA (2010)
19. AISC: Seismic Provisions for Structural Steel Buildings. ANSI/AISC 341-10. American Institute for Steel Construction, Chicago, IL (2010)

20. AISC: Prequalified Connections for Special and Intermediate Steel Moment Frames for Seismic Applications Including Supplement no. 1. ANSI/AISC, pp. 358–310. American Institute of Steel Construction, Inc., Chicago, IL (2010)
21. Lignos, D.G., Krawinkler, H.: Deterioration modeling of steel components in support of collapse prediction of steel moment frames under earthquake loading. J. Struct. Eng. **137**(11), 1291–1302 (2011)
22. PEER/ATC: Modeling and Acceptance Criteria for Seismic Design and Analysis of Tall Buildings. PEER/ATC 72-1. Pacific Earthquake Engineering Research Center, Redwood City, CA (2010)
23. Gupta, A., Krawinkler, H.: Seismic demands for the performance evaluation of steel moment resisting frame structures. Report No. 132. The John A. Blume Earthquake Engineering Center, Stanford University (1999)
24. Zareian, F., Medina, R.A.: A practical method for proper modeling of structural damping in inelastic plane structural systems. Comput. Struct. **88**(1), 45–53 (2010)
25. Simiu, E., Yeo, D.: Wind effects on structures: modern structural design for wind, 4th edn. Willey, NJ (2019)

Sea Breeze Occurrence on the Romanian Black Sea Coast

Crina Radu, Georgeta Bandoc, and Mircea Degeratu

Abstract This study aims to utilize the Sea Breeze Index (SBI) to identify sea breeze occurrences along the Romanian Black Sea coastline. The primary objective is to enhance the understanding of local climatology and validate empirical findings through a theoretical approach using the Copernicus European Regional Reanalysis (CERRA) database. By applying the SBI, we aim to delineate sea breeze patterns and their temporal variations. Preliminary results indicate that the SBI provides a more robust investigation and offers a comprehensive exploration of sea breeze dynamics. Sea breeze occurrences have been classified into four distinct types: blocked by offshore synoptic flow, sea breeze regime, embedded in onshore synoptic currents, and inhibited by a negative thermal gradient. The findings reveal a significant 70% increase in case detection compared to empirical methods, with over 80% of cases associated with sea breezes embedded in the synoptic current. The study also highlights discrepancies between the methods, providing insights into the strengths and limitations of each approach. Furthermore, it serves as a foundation for future research aimed at refining methodologies and improving weather forecasting accuracy.

C. Radu (✉)
University of Bucharest, Faculty of Geography, Bucharest, Romania

Eurowind Energy SRL, Wind Competence Center, Bucharest, Romania

G. Bandoc
University of Bucharest, Faculty of Geography, Bucharest, Romania

Romanian Association for Wind Engineering (ARIV), Bucharest, Romania

Academy of Romanian Scientists, Bucharest, Romania

M. Degeratu
Romanian Association for Wind Engineering (ARIV), Bucharest, Romania

Academy of Romanian Scientists, Bucharest, Romania

Department of Hydraulics and Environmental Protection Tehnical University of Civil Engineering of Bucharest, Bucharest, Romania

I. Calotescu, C. I. Coşoiu (eds.), *Proceedings of the Third National Conference on Wind Engineering*, Springer Proceedings in Earth and Environmental Sciences,
https://doi.org/10.1007/978-3-032-09671-5_5

Keywords Sea breeze · Sea breeze index (SBI) · Black Sea

1 Introduction

A sea breeze is a local circulation that occurs between two areas with differing atmospheric pressures due to two surfaces heating at different rates, causing movement in the direction of the thermal gradient. The earliest recorded description of the sea breeze is attributed to Aristotle in *Meteorologica* during ancient Greece [1]. Currently, numerous papers document significant studies on sea breeze circulation [1–7]. The sea breeze circulation also has a mesoscale impact on extensive coastal regions. In addition to influencing local weather, it can facilitate the recirculation of pollution and smog, as well as airborne insects and pollen spread [4]. Notably, it contributes to the generation of convective precipitation through the presence and development of the breeze front. This evolution of knowledge regarding sea breezes is illustrated in Fig. 1, beginning with Aristotle's description in *Meteorologica*. Since then, the study of sea breezes has proliferated, with the number of related scientific papers increasing each year, currently exceeding 3000 publications. Despite the extensive international research, studies on sea breezes in Romania remain limited, with only a few focused on the Black Sea coast [8–12].

This study is significant as it initiates research on sea breezes along the Black Sea coast, emphasizing the need for a regional understanding of this circulation. In this paper, we aim to identify instances of sea breeze occurrences along the Romanian coastline, enhance our understanding of local climatology, and delineate patterns and temporal variations of sea breezes throughout the year.

2 Data and Methods

2.1 Empirical Method

The empirical method used provides an initial, hands-on approach to better understand the sea breeze phenomenon through human analytical thinking before transitioning to a theoretical method based on exact mathematical formulas. This approach enhances comprehension of sea breeze patterns and mechanisms by direct observation and data interpretation. The empirical method unfolds in several stages, as explained hereinafter.

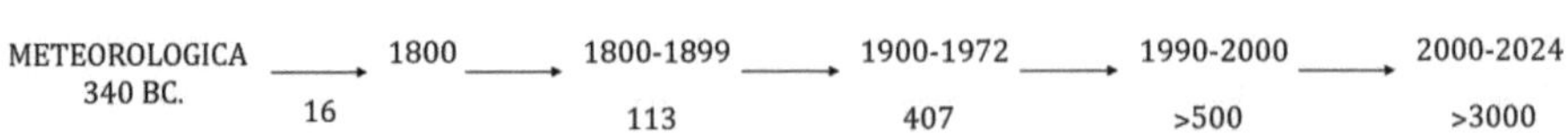

Fig. 1 Sea breeze knowledge evolution—from history to science

A comprehensive database of breeze cases was developed, drawing on a variety of data sources:

- *Meteorological Station Data:* Air temperature at 2 meters and water temperature readings from the Mangalia meteorological station (43.81°N, 28.58°E) were utilized to calculate the thermal gradient, which is crucial for identifying sea breeze conditions. Wind speed and direction at 10 meters, measured as 10-min averages and available at hourly intervals, were analyzed to define sea breeze patterns.
- *Satellite Imagery:* Visible-spectrum satellite images were employed to visually detect the sea breeze front, which serves as an important indicator of the breeze's presence and its behavior.
- *Numerical Forecast Models:* The Europe Swiss HD 4 × 4 model was utilized to ensure the accuracy of the extracted sea breeze cases by verifying the weather conditions favourable for their formation. The analysis included fields of temperature, geopotential at 500 hPa, mean sea level pressure, wind speed, and wind direction. This stage of analysis provided information about the atmospheric structure and wind dynamics.

Before interpretation, major features of the sea breeze system are identified as key parameters for analysis, including timing, duration, and velocity characteristics. These parameters are commonly used in sea breeze studies [13] and contribute to a better understanding of the sea breeze system, facilitating insights into its dynamics, behaviour, and impact on local weather patterns. This empirical method serves as a foundational step for researchers to filter, interpret, and assess sea breeze cases with clarity. It lays the groundwork for developing an automated theoretical model based on precise calculations, ensuring accurate detection and analysis of sea breezes.

To apply the empirical method for extracting sea breeze cases, the initial step involved analyzing meteorological data related to wind and temperature gradients. We focused on identifying days with a thermal gradient of at least 2° and with winds with an offshore direction. For excluding those with predominant synoptic winds we check the pressure fields and synoptic context ensuring that the selected cases were indeed driven by local sea breeze dynamics. Additionally, satellite imagery was extremely useful in confirming the presence of the sea breeze front, as it allowed for the visual identification of linear cumulus formations typically associated with the sea breeze.

2.2 Theoretical Method

For the theoretical method, the Sea Breeze Index (SBI) from Eq. 1 is applied. SBI ($m^2\ s^{-2}\ K^{-1}$), where U is the wind speed ($m\ s^{-1}$) and ΔT is the thermal gradient (°K). This equation was developed by Walsh [14] and Simpson [4], and by other researchers subsequently, such as Frysinger [15]. To define the equilibrium state when the synoptic offshore wind balances the thermal gradient flow, Eq. 2 is applied where h represents the thickness of the air mass that exchanges heat with the surface, and g is considered constant over the small vertical extent of the sea breeze. Eq. 1 was applied to CERRA (Copernicus European Regional Reanalysis) which has a three-hour frequency (UTC time) at near surface and 50 m levels for complete data series from 2 grid points with 5.5 km horizontal resolution located at 43.81,28.55 (onshore point—near west from Mangalia Meteorological station) and 43.81,28.65 (offshore point).

$$SBI = \mp\frac{U^2}{\Delta T} \tag{1}$$

$$U^2 = \left(\frac{2gh}{T}\right)(T_{air} - T_{sea}) = \left(\frac{2gh}{T}\right)\Delta T \tag{2}$$

$$\frac{U^2}{\Delta T} = \frac{2gh}{T} = SBI \tag{3}$$

The critical value of SBI depends on the conditions of each coast, so it can vary. Values from 2.7 ($m^2\ s^{-2}\ K^{-1}$) [15] to 10 ($m^2\ s^{-2}\ K^{-1}$) [16] were identified. A critical value established by Biggs and Graves [17] for the lake breeze index established a transition zone between (3.0 ± 0.2) ($m^2\ s^{-2}\ K^{-1}$), which turned out to be a procedure with an accuracy of 97% in the test period.

To determine the critical Sea Breeze Index, Bernoulli's equation (Eq. 4) is used as a relationship written between specific energies referenced to fluid weight which is applied between points 1 and 2 located on a streamline within the atmospheric boundary layer at a height H above the land (point 1) and above the sea (point 2) denoted as T_1 and T_2. These specific energies are influenced by the air temperatures above the land T_1 and above the sea T_2.

$$\frac{z_1 + \frac{p_1}{\rho g} + \frac{u_1^2}{2g}}{T_1} = \frac{z_2 + \frac{p_2}{\rho g} + \frac{u_2^2}{2g}}{T_2} \tag{4}$$

$$\frac{2gH + U^2}{T_{land}} = \frac{2gH}{T_{ov\ sea}} \tag{5}$$

$$U^2 = 2gH\frac{\Delta T}{T} \tag{6}$$

A critical value was established at 3.4 ($m^2\ s^{-2}\ K^{-1}$) for the Romanian coast, which represents the critical sea breeze index (at equilibrium) expressed in Eq. 7.

$$SBI_{cr} = \frac{2gH}{T} \tag{7}$$

Then, the effective sea breeze index SBI is compared to the critical sea breeze index SBI_{cr}. If $SBI < SBI_{cr}$, a sea breeze regime is identified. If $SBI > SBI_{cr}$, a sea breeze blocked by the offshore synoptic current is identified.

3 Results

Many studies on sea breezes primarily focus on those occurring during the summer or the warm season. However, fewer studies consider sea breezes throughout the entire year, as they are often believed to be absent in the cold season. In reality, while less frequent in winter, sea breezes do occur. Under certain conditions, which are becoming increasingly common due to global warming, thermal gradients can be strong enough to produce intense sea breezes even in colder months. This research also examines sea breezes during the cold season, thereby enriching our overall understanding of sea breeze dynamics across all seasons.

3.1 Empirical Results

Initially, we employed to use an empirical method to gain a closer understanding of how sea breezes behave. The intent behind this approach was to rely on human reasoning and interpretation to closely analyze the patterns and characteristics of sea breezes. Filters were applied based on observational insights, using a 10-minute averaging period, which allowed for a deeper understanding of the phenomena.

The first results provide statistics on sea breeze occurrences based on the empirical method, categorizing them into pure and embedded sea breeze cases. Because a single sea breeze event can span multiple days, or multiple events can occur within a single day, a statistic based on the number of days with sea breeze occurrences (as shown in Tables 1 and 2) is more suitable for characterizing the long-term frequency of sea breeze. According to Table 1 the empirical method shows that the total number of days with sea breeze events in 2018 adds up to 122. This includes 37 days with pure sea breeze and 85 days with embedded sea breeze. During the cold season, the empirical method identified 9 days with pure sea breeze and 11 days with embedded sea breeze, totalling 20 days of sea breeze. In the warm season, there were 28 days with pure sea breeze and 74 days with embedded sea breeze, adding up to 102 days.

Based on empirical methods, we considered 2 types of sea breeze which are represented as distinct cases for pure sea breeze (Fig. 2) and embedded sea breeze (Fig. 3). As depicted in Fig. 2, on June 22, a significant shift in wind direction occurred around 10:00 UTC, with the wind turning from the west to east—indicating the onset of a sea breeze originating from the Black Sea towards the Romanian coast. Concurrently, the wind speed increased, peaking at 5 m/s around

Table 1 Number of days with sea breeze types occurrence based on empirical method

Period	Days with pure sea breeze	Days with embedded sea breeze	Total sea breeze cases
Cold season	9	11	20
Warm season	28	74	102
2018	37	85	122

Table 2 Number of days with sea breeze types occurrence based on theoretical method

Period	Days with blocked SB (C1a)	Days with Pure SB (C1b)	Days with Inhibited SB (C2)	Days with Inhibited SB (C3)	Days with Embedded SB (C4)
Cold season	13	39	140	121	50
Warm season	16	50	142	126	131
2018	29	89	282	247	181

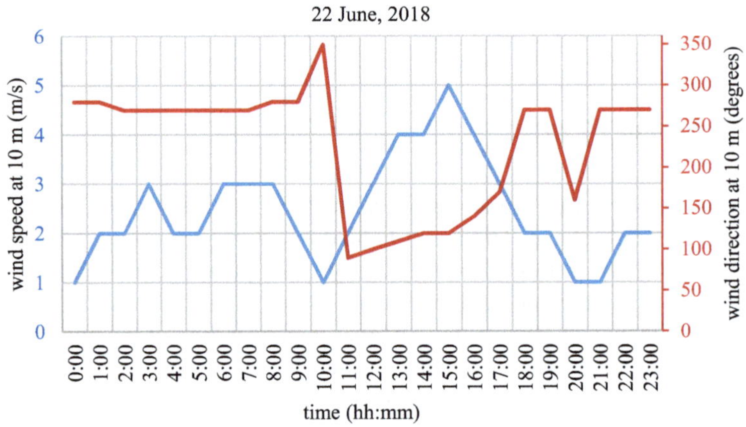

Fig. 2 Sea breeze dynamics: line chart correlation between wind speed and direction throughout the day for a pure sea breeze type

15:00 UTC. This significant increase in wind speed is, in general, indicator of the coexistence of the sea breeze with synoptic flow.

Figure 3 illustrates a pure sea breeze occurrence, devoid of the influence of a synoptic flow. The sea breeze initiates in the morning hours, with wind speeds gradually increasing and reaching a peak of 9 m/s in the early afternoon when the thermal gradient is maximum. The absence of interaction with synoptic flow in this case results in a more gradual rise and fall of wind speed, with no abrupt shifts in direction.

In both cases, the onset of the breeze is triggered by the differential heating between the land and the sea, which is more pronounced during midday. The maximum wind speed in the combined breeze scenario is significantly higher due to the additive effect of synoptic flow, whereas the pure breeze exhibits a more

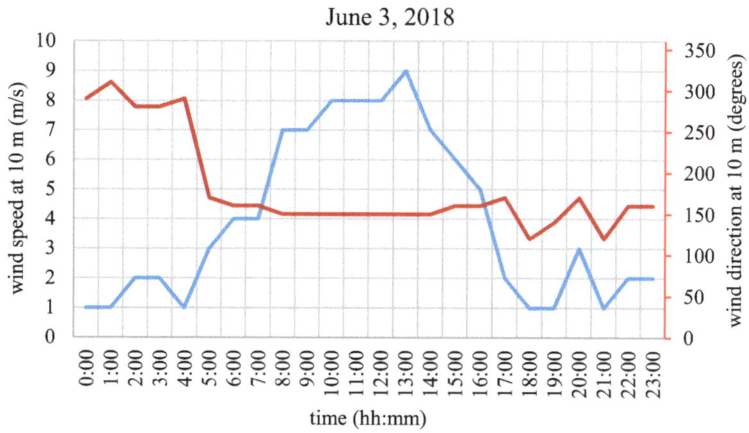

Fig. 3 Sea breeze dynamics: line chart correlation between wind speed and direction throughout the day for an embedded sea breeze type

moderate peak. The embedded sea breeze type diminishes as synoptic influences wane, while in the pure breeze case, it subsides as the land-sea temperature difference decreases by evening.

3.2 *Theoretical Results*

To validate and refine the results from the empirical method we applied a more technical approach using the Sea Breeze Index (SBI). This method provided a systematic and objective way to quantify sea breeze occurrences, ensuring that the initial empirical observations could be confirmed and expanded upon. By incorporating the SBI, we achieved a higher level of precision in identifying sea breeze events and eliminate potential biases that might have emerged from human-driven analysis. This combination of empirical and technical methods allowed for a more robust and comprehensive understanding of sea breeze dynamics.

The statistical analysis confirms the efficiency of the method as the total number of days with sea breeze (both pure and embedded) adds up to 270. This indicates that the technical method identified 148% more days with sea breeze compared to the empirical method.

The SBI quadrant plot for 2018 is shown in Fig. 4, categorizing the occurrences of sea breeze events throughout the year based on their types. Data was recorded every 3 h, with each case representing the SBI value at a specific time step. Pure sea breezes, highlighted in red, predominantly occur in quadrant C1a, where the synoptic offshore flow is dominant but weak enough to allow the formation of sea breezes. The distribution shows that most cases cluster near the origin, indicating a

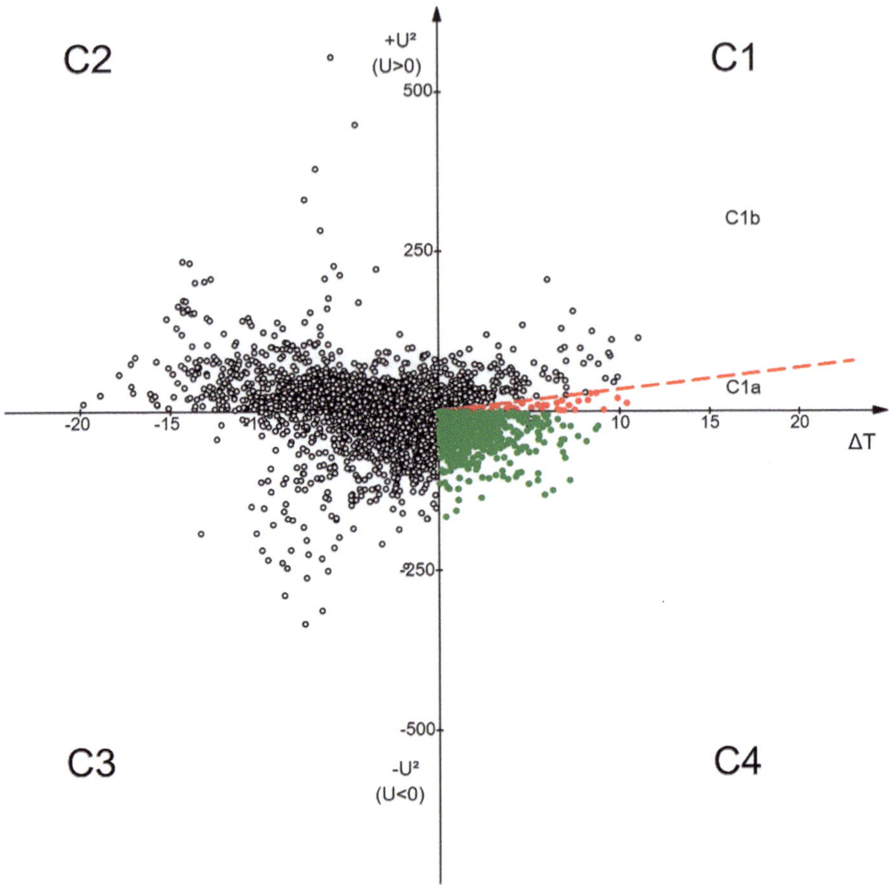

Fig. 4 SBI plot for 2018 cases. C1 = offshore synoptic flow; C1a = SB regime and pure SB with SBI < 3.4 (m^2 s^{-2} K^{-1}); C1b = SB blocked by synoptic flow; C2 = offshore synoptic flow with SB inhibited by thermal gradient; C3 = onshore synoptic flow with SB inhibited by thermal gradient; C4 = onshore synoptic flow, possibly with SB embedded

balance between onshore and offshore synoptic flows, with inhibited or embedded sea breezes occurring more frequently than pure sea breezes.

The relationship between synoptic gradient wind and temperature gradient wind in each type of sea breeze is detailed below. U represents the wind due to the synoptic pressure gradient (baric gradient), while ΔT leads to the formation of a current (wind) due to the temperature difference (temperature gradient wind), which constitutes the sea breeze.

- If U is positive, it results in a synoptic offshore wind (from land to sea).
- If ΔT is positive, it results in a temperature gradient wind from sea to land (sea breeze).

Therefore:

In the first quadrant (C1), where U is positive and ΔT is positive, the synoptic current (wind) opposes the temperature gradient wind, resulting in a sea breeze. In C1a the offshore synoptic gradient wind is less strong than the temperature gradient wind, which is linked to the sea breeze. This difference in strength allows the sea breeze to dominate, resulting in the formation of a pure sea breeze. In C1b within the first quadrant, the offshore synoptic gradient wind is stronger than or at least equal to the temperature gradient wind (associated with the sea breeze), thus blocking the sea breeze.

In C2, where U is positive and ΔT is negative, the synoptic offshore wind (from land to sea) is dominant. However, a negative ΔT indicates that the temperature gradient favors a wind from land to sea (land breeze). In this case, both the synoptic wind and the temperature gradient wind move in the same direction, which leads to the inhibition of the sea breeze since the circulation is predominantly offshore. Essentially, the conditions in C2 favor a strong offshore synoptic flow, with the sea breeze being suppressed by the thermal gradient.

In C3, where U and ΔT are both negative, the synoptic wind is onshore (from sea to land), while negative ΔT indicates a temperature gradient wind moving from land to sea, corresponding to a land breeze. In this case, the synoptic wind and the temperature gradient wind oppose each other inhibiting the sea breeze due to this directional conflict. In this quadrant, the onshore synoptic flow is strong enough to prevent the land breeze but not strong enough to allow a sea breeze to develop. Thus, in C3, there is an onshore synoptic flow, but no significant contribution from the sea breeze, which is inhibited by the temperature gradient.

In C4, where U is negative and ΔT is positive, the onshore synoptic gradient wind from the sea to land aligns with the wind caused by the temperature gradient (the sea breeze). As a result, these two winds combine, integrating the sea breeze within the onshore wind (embedded sea breeze).

Figure 5 shows the monthly average duration of both pure sea breezes (marked in red) and embedded sea breezes (marked in green) throughout the year. The visual representation indicates key patterns in the distribution and duration of sea breezes.

3.2.1 Cold Season (October to March)

Embedded Sea Breezes

- Embedded sea breezes are predominantly observed during the cold season, particularly in October, November, and December, where they have long durations.
- In December, embedded sea breezes extend from approximately 7:00 AM until late in the afternoon (close to 3:00 PM).
- In January, February, and March, embedded sea breezes are present but occur for shorter periods, typically in late morning or early afternoon (e.g., in January from 8:00 AM to around 2:00 PM).

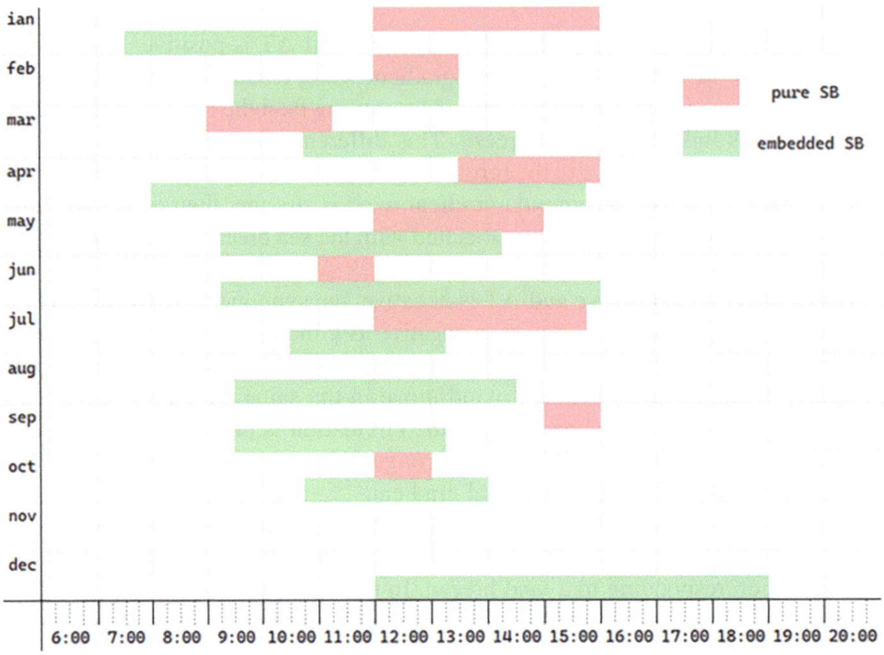

Fig. 5 Monthly average duration of Sea Breezes

Pure Sea Breezes

- Pure sea breezes are much less frequent and occur for shorter durations.
- In February and March, pure sea breezes typically appear only in the afternoon, usually for a few hours (e.g., February from around 1:00 PM to 3:00 PM).
- Pure sea breezes are completely absent in December and November but appear more sporadically in October.

Key Observations

- Embedded sea breezes dominate and occur for longer periods, generally starting in the morning and lasting until late afternoon.
- Pure sea breezes, when they occur, are short-lived and typically appear in the afternoon during the transition from winter to spring.

3.2.2 Warm Season (April to September)

Embedded Sea Breezes

- Embedded sea breezes remain common throughout the warm season, especially in April, May, and the peak summer months of June, July, and August.
- In June and July, embedded sea breezes last from early morning to late afternoon or early evening, with durations extending from 8:00 AM to 5:00 PM or later.

- September shows a slight decrease in the length of embedded sea breezes compared to the peak summer months, but they are still significant.

Pure Sea Breezes

- Pure sea breezes are more frequent during the warm season, particularly in summer months, where they tend to occur in the afternoon, even for shorter periods.
- In July, pure sea breezes typically occur from approximately 2:00 PM to 5:00 PM, while in August, they are shorter, lasting around 2 h.
- May also sees some occurrences of pure sea breezes, although they are less prominent compared to embedded breezes.

Key Observations

- Embedded sea breezes are dominant, especially during the summer months, with long durations throughout the day.
- Pure sea breezes are more frequent during the warm season but typically occur in the afternoon and last for shorter periods.

This seasonal structure highlights that embedded sea breezes are key feature of both the cold and warm seasons, while pure sea breezes become more prominent during the summer albeit with shorter durations.

4 Conclusion

The main advantage of the technical method is its capacity to minimize human errors that arising from the subjective interpretation of each case.

The values of sea breeze cases identified using the empirical method are significantly lower than those identified using the theoretical method, as the empirical method tends to detect fewer sea breeze occurrences, particularly embedded cases. This demonstrates the more conservative approach for the empirical method in identifying sea breeze events compared to the theoretical method.

The results obtained through the technical method are more consistent than those from the empirical method and highlight a noticeable concentration of pure and embedded sea breezes.

Embedded sea breezes tend to occur more frequently and have longer durations throughout the year, particularly in the summer and fall months. In contrast, pure sea breezes are less common, and when they do occur, they typically have shorter durations, especially outside the summer period. This underscores the strong influence of synoptic conditions in supporting embedded sea breezes during most months.

These conclusions emphasize the importance of detailed and objective analysis in better understanding sea breeze dynamics, which can contribute to improved management and planning of activities in coastal areas.

Furthermore, this research lays the groundwork for further investigation and a better understanding of sea breezes along the Romanian Black Sea coast. In our future research, we will increase the number of analyzed parameters and extend and apply the technical method to develop a climatology of sea breezes along the Romanian Black Sea coast, utilizing long-term meteorological data. This approach will enhance the existing literature and contribute to more accurate weather forecasts, which are essential for climate assessments specific to the Romanian coastal environment.

References

1. Miller, S.T.K., Keim, B.D., Talbot, R.W., Mao, H.: Sea breeze: structure, forecasting, and impacts. Rev. Geophys. **41**(3), 1 (2003)
2. Baralt, G.L., Brown, R.A.: The Land and Sea Breeze: an Annotated Bibliography. Contract No. AF19(604)-7259, Final Report (1960–65). Department of Geophysical Sciences, University of Chicago, Chicago (1965)
3. Jehn, K.H.: A Sea Breeze Bibliography, 1664–1972, vol. 37. Atmospheric Science Group, University of Texas (1973)
4. Simpson, J.E.: Sea Breeze and Local Winds, p. 234. Cambridge University Press (1994)
5. Crosman, E.T., Horel, J.D.: Sea and Lake breezes: a review of numerical studies. Bound.-Layer Meteorol. **137**(1), 1–29 (2010)
6. Guo, Z., Cao, L., Guo, J., Shang, M., Guo, X.: On three types of sea breeze in Qingdao of East China: an observational analysis. Front. Environ. Sci. **11**, 1188952 (2023)
7. Shang, M., Cao, L., Guo, J., Guo, Z., Liu, L., Zhong, S.: Influence of pure breeze on urban heat Island in Tianjian, China: a perspective from multiple meteorological observations. Atmospheric Reserch. **304**, 107 408 (2024)
8. Kolev, I., Parvanov, O., Kaprielov, B., Donev, E., Ivanov, D.: Lidar observations of sea-breeze and land-breeze aerosol structure on the Black Sea. J. Appl. Meteorol. **37**(10), 982–995 (1998)
9. Skakalova, T.S., Savov, P.B., Grigorov, I.V., Kolev, I.N.: Lidar observation of breeze structure during the transition periods at the southern Bulgarian Black Sea coast. Atmos. Environ. **37**(3), 299–311 (2003)
10. Efimov, V.V., Krupin, A.V.: Breeze circulation in the Black Sea region. Russ. Meteorol. Hydrol. **41**(4), 240–246 (2016)
11. Savov, P., Kolev, N., Kolarova, M., Batchvarova, E., Barantiev, D.: Aerosols, Ozone and CO2 under Sea-Breeze Conditions at a Black Sea Coastal Site. (2019)
12. Radu, C., Bandoc, G., Degeratu, M.: Advancing Understanding of the Sea Breeze for the Romanian Black Sea Coast: A Comparative Analysis Using the SBI Index, EMS2024-462. Copernicus Meetings (2024)
13. Clark, I.W.: A Three-Dimensional Numerical Model of the Sea Breeze for the Plymouth Region. Thesis, University of Plymouth (1986)
14. Walsh, J.E.: Sea breeze theory and applications. J. Atmos. Sci. **31**, 2012–2026 (1974)
15. Frysinger, J.R., Lindner, B.L., Brueske, S.L.: A statistical sea-breeze prediction algorithm for Charleston, South Carolina. Weather Forecast. **18**(4), 614–625 (2003)
16. Lyons, W.A.: The climatology and prediction of the Chicago lake breeze. J. Appl. Meteorol. **11**, 1259–1270 (1972)
17. Biggs, W.G., Graves, M.E.: A lake breeze index. J. Appl. Meteorol. **1**(4), 474–480 (1962)

Damping Estimation of a Telecommunication Lattice Tower from Full-Scale Ambient Vibration Measurements

Ileana Calotescu, Ioana-Claudia Tomescu, and Daniel Bîtcă

Abstract Free-standing lattice towers are lightweight modular structures used in the field of telecommunications to support antenna needed for signal transmission. Due to their purpose, lattice towers reach considerable heights which, together with their reduced mass may render the structure sensitive to dynamic wind loading. This paper presents the estimation of damping of a 50 m high triangular telecommunication lattice tower from full-scale acceleration measurements. The tower is located in Afumați, Ilfov County, Romania and is equipped with a monitoring system comprising an ultrasonic anemometer and a triaxial accelerometer which were installed in September 2023. Full scale wind velocity and acceleration stationary records are used in order to analyze the wind-induced behavior of the tower. The modal parameters are estimated by making use of the Frequency Domain Decomposition method whereas the damping is evaluated based on the logarithmic decrement technique.

Keywords Damping · Frequency domain decomposition · Lattice towers · Wind

1 Introduction

Telecommunication lattice towers are used in fields of radio and television broadcasting to support necessary equipment to certain heights. Due to their slender geometry and low structural damping, these towers are particularly susceptible to wind-induced vibrations, which can lead to serviceability issues or even structural fatigue over time [1, 2]. Thus, damping is a critical factor in their design [3].

The purpose of this paper is to identify the modal parameters (particularly damping ratios) of a 50-meter-high telecommunication lattice tower using full-scale acceleration measurements and to validate these results with code-based values.

I. Calotescu (✉) · I.-C. Tomescu · D. Bîtcă
Technical University of Civil Engineering Bucharest, Bucharest, Romania
e-mail: ileana.calotescu@utcb.ro

I. Calotescu, C. I. Coşoiu (eds.), *Proceedings of the Third National Conference on Wind Engineering*, Springer Proceedings in Earth and Environmental Sciences,
https://doi.org/10.1007/978-3-032-09671-5_6

Full-scale tests performed on free-standing lattice towers show a wide range of values of the damping ratios. Balio et al. [4] investigated the behavior of a 100 m high lattice tower through full scale measurement of wind velocity and acceleration response and obtained damping rations between 0.3% and 0.5%. The authors also acknowledged a modest dependence of the damping ratio with mean wind velocity which might be related to aerodynamic damping. Similarly, Holmes et al. [5] investigated the wind induced behavior of a 104 m tall lattice tower used for radio and television transmissions by means of acceleration measurements. A free vibration analysis was carried out to identify the dynamic characteristics of the tower. The total damping ratio was estimated at 0.7%. A full-scale measurement program was also conducted by Glanville and Kwok [6] on a 67 m communication lattice tower in Sydney. Force vibration tests were performed during still atmospheric conditions to investigate the dynamic parameters of the tower. The damping ratio was found to be approximately 1%. The authors also stated that the presence of ancillaries which constituted 5% of the tower's projected area had a negligible effect on the dynamic wind-induced response of the tower. Recently, Garcia et al. [7] obtained values of damping ratios in the range of 0.20%–0.3% from full scale acceleration measurements of a 58 m tall antennae lattice tower. The authors explain the low values by the fact that records corresponding to low wind velocities were used in the evaluation of damping ratio (less than 10 m/s).

Modal identification of structures and estimation of damping from structural response time histories include both frequency domain methods (e.g. Frequency Domain Decomposition method) as well as time domain methods (e.g. Random Decrement technique) [8, 9]. Some key challenges in estimating damping from full-scale field measurements involve distinguishing between closely spaced modes of vibration, dealing with non-stationary excitation caused by variable wind loads, and separating structural damping from aerodynamic damping. Additionally, the low signal-to-noise ratio in ambient vibration data and the influence of ancillary equipment on the dynamic response further complicate accurate damping estimation. These challenges necessitate robust signal processing techniques and careful experimental design to obtain reliable damping values for structural assessment and design validation. To address these challenges, his study makes use of the Frequency Domain Decomposition (FDD) technique [10, 11] in which the modal identification is carried out based on the ambient structural responses. The required conditions for the application of this technique include stationarity of data which may be ensured by stationarity tests [12] and a high quality of the response PSD (Power Spectral Density) function which may be ensured by long enough data and well separated modes of vibration [9]. The damping is estimated by means of the logarithmic decrement technique. Finally, results are cross validated against those obtained from Finite Element Analysis (FEA) and the Romanian wind code [13].

Section 2 presents the geometric characteristics of the monitored tower, and the dynamic parameters obtained from FEA. In Sect. 3, the monitoring system is introduced and the full-scale measurements used for the identification of the dynamic characteristics are presented. The estimation of damping is described in Sect. 4 whereas Sect. 5 concludes the paper.

2 The Monitored Tower

The monitored structure is a 50 m high triangular telecommunication lattice located in Afumați, Romania [14] (see Fig. 1). The tower is divided into 10 sections and is tapered up to 40 m with a 3.72° inclination with respect to the vertical. From 40 m up to 50 m the tower has parallel legs. The lower 5 sections are made up of inverted V-bracing systems whereas the upper 5 sections are made up of N-bracing systems. The distance between leg members is 7.50 m at the base and 2.30 m at the top. All members have circular hollow cross-sections and are connected though gusset plates and high-strength bolts. Along the tower height there are two resting platforms at 15 m and 27.50 m as well as two working platforms at 40 m and 47.50 m. The mass of the ancillary elements is approximately 4.5 tons, and the total mass of the structure and ancillaries is 13.7 tons.

The dynamic characteristics of the tower were obtained from the Finite Element Analysis (FEA). All connections were considered as pinned resulting in elements subjected to only axial forces. The platforms and antennas were modelled as concentrated forces acting on the leg members positioned at heights corresponding to each element. The self-weight of the ladder and cables was evenly distributed along the height of the tower. Table 1 summarizes the frequency and shape of the first six modes of vibration as obtained from FEA. Due to the symmetry of the tower, the flexural modes occur in pairs with vibration parallel and perpendicular to one face characterized by identical natural frequencies.

3 Full-Scale Measurements

The tower is equipped with an ultrasonic anemometer compact (UA), a triaxial accelerometer (A50) and a video monitoring system (CCTV1–3) as denoted in Fig. 2. The anemometer is installed on a 1.5 m rod above the tower so that interference

Fig. 1 (**a**) The monitored tower (**b**) Tower location

Table 1 Modes of vibration of the monitored tower from FEA

Mode	1	2	3	4	5	6
Freq. (Hz)	1.47	1.47	3.13	4.76	4.76	5.95
Modal shape					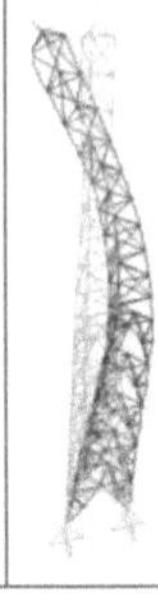	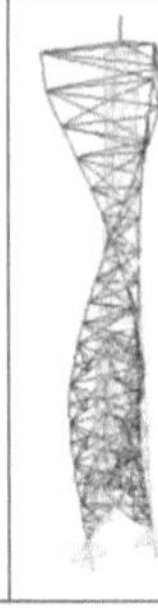

from the tower elements and ancillaries is limited. The accelerometer is aligned with its y-axis parallel to face 1–3 of the tower while x-axis is perpendicular to that face and z-axis is in the vertical direction (see Fig. 2). It is installed at 50 m on tower Leg 1. The sampling rate of the anemometer is 10 Hz whereas the sampling rate of the accelerometer is 125 Hz. The sensors are connected to a datalogger and transmit the data in real time. The monitoring system has been continuously functioning since September 2023.

In this paper, 20 stationary 1-hr velocity records were extracted from the database and the corresponding acceleration records were used in the FDD technique to estimate the modal parameters of the tower and the damping coefficients. A stationary record is a record whose statistical properties such as mean and variance do not change significantly over time. As such, the stationarity of each velocity record was established by limiting the relative error between the moving average mean evaluated on a 10-min window and the 1-hr mean. The error was limited to 15%. Furthermore, to ensure the stationarity of the corresponding acceleration records, the run test [12] was applied.

Considering a sample acceleration record, the run test is performed by dividing the sample record into N equal time intervals. In each interval, the data may be considered independent. Next, the mean square value is computed for each interval and the sequence of mean square values is tested for the presence of underlaying trends or variations. The statistical parameters of the selected 1-hr velocity records, i.e. the peak wind speed v_{p}, the mean speed v_{m60}, the mean wind direction d_{m60}, the skewness s_{v60} and the kurtosis k_{v60} of wind speed together with the peak (a_{max}) and root mean square of acceleration (a_{RMS}) on x and y directions are shown in Table 2. The 2023-12-23 11-30-44 record is analyzed in detail hereinafter.

Figure 3 shows the wind speed and direction of time-histories together with their corresponding Probability Density Functions (PDFs) for the wind record measured on December 23rd 2023 (2023-12-23 11–30-44). The peak wind speed was $v_{\mathrm{p}} = 20.80$ m/s while the 1-hr mean resulted $v_{\mathrm{m60}} = 12.02$ m/s (represented by the black dotted line in Fig. 3). The skewness and kurtosis resulted $s_{60} = 0$ and

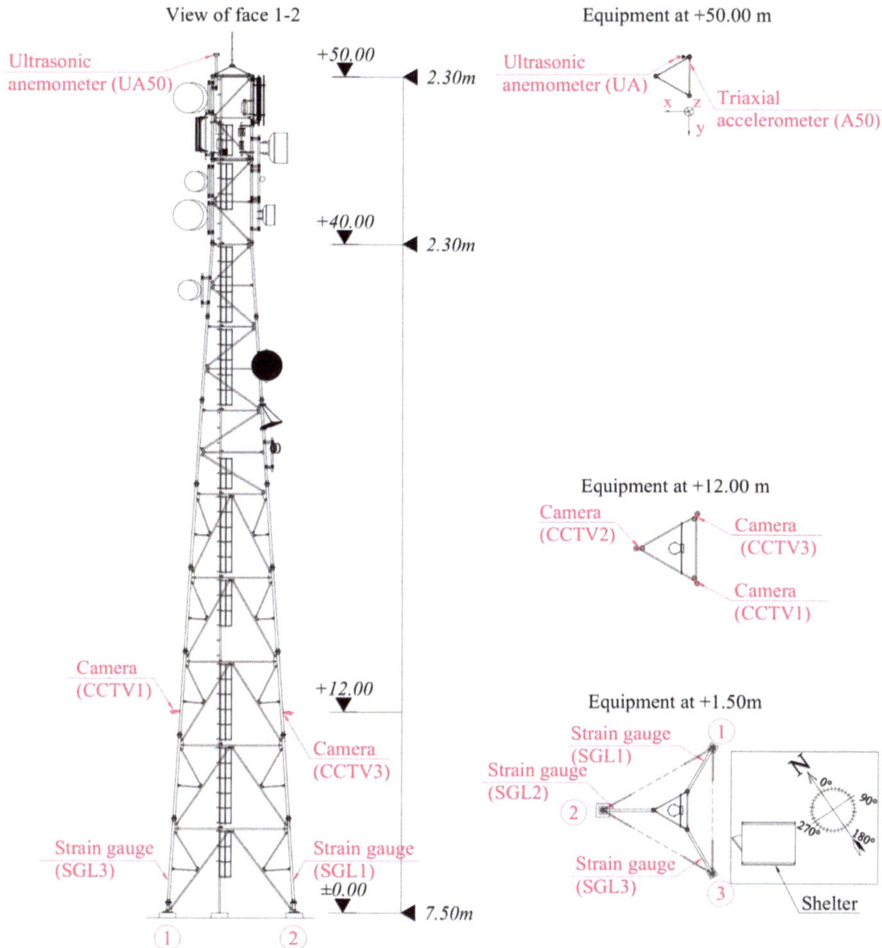

Fig. 2 Sketch of the lattice tower and position of installed sensors

$k_{60} = 2.62$ respectively which are close to the typical Gaussian distribution ($s = 0$ and $k = 3$).

The average wind direction was 264° which is approximately aligned to the y direction of the accelerometer (Face 1–3 in Fig. 2). During the 1-hr interval, the wind direction did not change considerably, ranging from 237° to 310°, the coefficient of variation (COV) resulting 0.03. The temperature was also approximately constant throughout the 1-hr interval, with an average of 6.4 °C. Figure 4 shows the variation of the wind direction and corresponding PDF for the December 23rd 2023 record.

The Power Spectral Density functions (PSDs) of the ensemble of the 20 stationary velocity records (Table 2) together with the average value (thick line) are shown in Fig. 5a. The slope of the curve $n^{-5/3}$ related to the inertial sub range

Table 2 Stationary records and corresponding statistical parameters

Record (date and time)	v_p(m/s)	v_{m60}(m/s)	d_{m60} (°)	s_{v60}	k_{v60}	a_{max} (cm/s^2)	a_{RMS} (cm/s^2) x-dir	y-dir
2023-11-23 19-55-56	14.34	7.65	249	0.16	3.17	11.44	1.16	1.14
2023-11-24 01-02-06	15.02	7.95	257	0.22	3.13	10.03	1.07	1.10
2023-11-24 10-46-50	10.56	6.06	265	0.07	2.82	4.95	0.51	0.58
2023-11-24 16-05-19	9.26	5.40	250	0.04	3.08	3.06	0.45	0.49
2023-11-27 10-58-35	11.04	6.62	273	0.00	2.75	5.00	0.58	0.63
2023-12-17 05-10-19	6.28	4.55	259	−0.20	2.96	1.15	0.20	0.22
2023-12-21 00-13-58	8.94	5.59	265	0.13	3.16	2.30	0.31	0.36
2023-12-22 20-25-46	17.56	10.78	262	0.06	2.67	17.29	2.02	2.48
2023-12-22 21-28-48	14.6	8.74	262	0.25	2.75	9.84	1.09	1.23
2023-12-23 11-30-44	20.8	12.02	267	0.00	2.62	21.50	2.31	3.00
2023-12-23 21-37-41	12.64	7.23	273	−0.02	2.70	5.77	0.69	0.80
2023-12-25 10-23-00	14.2	7.39	255	0.34	2.87	8.73	0.99	1.05
2024-01-03 21-08-03	13.96	8.27	266	0.01	2.61	6.87	1.00	1.05
2024-02-05 13-58-19	18.46	11.06	261	0.09	2.86	22.96	2.55	3.16
2024-02-05 22-49-53	12.7	8.13	260	0.23	2.86	6.90	0.90	0.98
2024-02-06 14-00-03	17.44	10.76	251	−0.17	2.74	18.88	2.20	2.71
2024-02-06 14-35-25	16.86	10.44	251	−0.13	2.61	20.23	2.10	2.55
2024-04-02 17-45-04	23.4	14.63	266	−0.08	2.57	37.81	3.94	5.24
2024-04-02 18-22-08	22.9	13.58	265	−0.08	2.50	39.99	3.70	4.90
2024-04-16 16-30-37	14.3	7.14	245	0.01	3.10	8.62	0.86	0.88

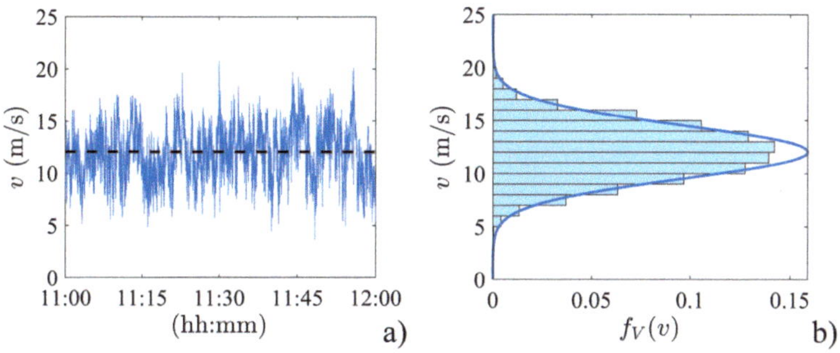

Fig. 3 December 23rd 2023 record: (**a**) wind speed and (**b**) PDF of wind speed

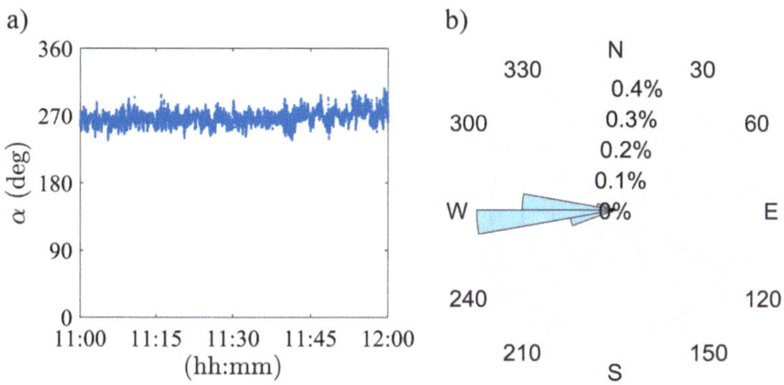

Fig. 4 December 23rd 2023 record: (**a**) wind direction and (**b**) PDF of wind direction

is also plotted, showing a good fit with the measured data. Figure 5b shows the normalized PSDs corresponding to the 20 records together with their average (thick line) compared to the PSD provide by the Romanian wind loading code, CR 1-1-4/2012 [13] which is based on the model proposed by Solari and Piccardo [15]. A good fit was obtained between the full-scale measurements and the code-based recommendations.

Finally, Fig. 6 shows the wind-induced acceleration time-histories corresponding to the December 23rd 2023 record along the x and y directions of the accelerometer. The maximum resultant acceleration was $a_{\max} = 21$ cm/s^2. The power spectral density functions (PSDs) of acceleration signals were obtained using Welch's averaged periodogram method. The first flexural modes with spectral peaks at 1.46 Hz on both directions are clearly identified in the acceleration PSDs (Fig. 6b) and they match well the corresponding values obtained from FEA of the tower (Table 1). As may be seen, higher order modes do not match the corresponding FEA predictions. This may be caused by various reasons including simplifications in the FEA modeling, limited sensor resolution for higher modes, and the influence of non-structural components on the tower's dynamic behavior.

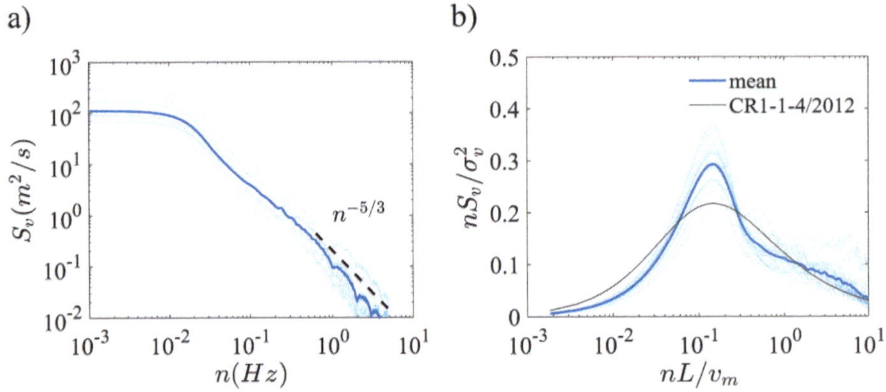

Fig. 5 (**a**) PSDs and (**b**) normalized PSDs of the ensemble of stationary wind records

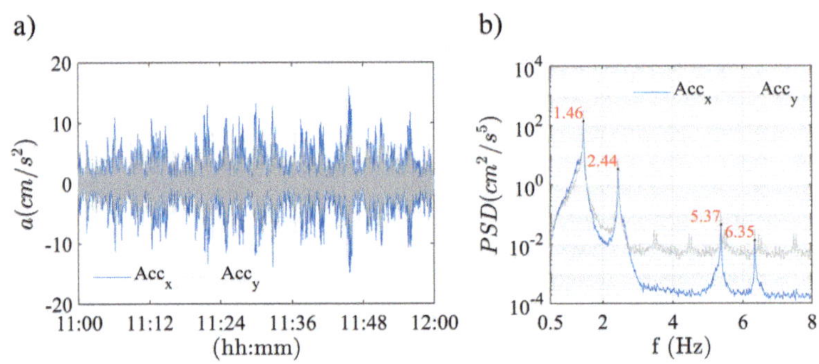

Fig. 6 December 23rd 2023 record: (**a**) acceleration and (**b**) PSD of acceleration

4 Estimation of Damping

Damping represents the ability of a structural system to dissipate vibrational energy, and it mainly consists of three components: structural damping, aerodynamic damping and damping due to special devices.

Structural damping depends on several parameters such as material, structural configuration, connections of elements as well as frequency of vibration. In general, structural damping is assumed to be of viscous-type and is expressed as the logarithmic decrement of damping (δ_S) or the damping ratio (ξ_S) which may be obtained from experimental tests using various methods [8, 9]. The Romanian wind code [13] provides damping ratios for steel towers according to the connection types. For lattice towers with elements connected by high strength bolts the recommended value is $\xi_S = 0.5\%$. Aerodynamic damping is caused by the relative velocity of the structure to that of the air flow. For lightweight structures, aerodynamic damping in the along-wind direction is considerably larger than structural damping leading to a reduction of wind-induced response.

In this paper, the modal parameters of the 50 m tower are estimated by making use of the Frequency Domain Decomposition (FDD) technique [10, 11] in which the modal identification is carried out based on the acceleration measurements. Prior to analysis, the acceleration signals were filtered using a both a lowpass and subsequently a highpass Butterworth filter, the cut-off frequencies being 0.5 Hz and 20 Hz respectively. The number of data points used in the estimation of the acceleration PSDs are important in the accurate estimation of damping as pointed out by Tamura et al. [9]. As such, for the purpose of this analysis, 2^{16} data points were used.

Let us consider a sample acceleration signal measured by the monitoring system as exemplified in Eq. 1, where $a_1(t)$ and $a_2(t)$ represent the acceleration signals measured at heights 0 and 50 m respectively:

$$A(t) = \begin{bmatrix} a_1(t) \\ a_2(t) \end{bmatrix} \tag{1}$$

The modal identification algorithm employed in the FFD technique involves the following steps [10, 11]:

1. estimate the output Spectral Density Matrix (SDM). In case only one accelerometer is used, the SDM may be expressed as:

$$G(f) = \begin{bmatrix} a_1(f) \\ a_2(f) \end{bmatrix} \left[a_1^*(f) a_2^*(f) \right] = \begin{bmatrix} a_1^2(f) & a_1(f) \cdot a_2^*(f) \\ a_2(f) \cdot a_1^*(f) & a_2^2(f) \end{bmatrix} \tag{2}$$

where $a_1(f)$and $a_2(f)$are determined by taking the Fast Fourier Transform (FFT) of the acceleration time signals $a_1(t)$ and $a_2(t)$ and $a_1^*(f)$ and $a_2^*(f)$denote the complex conjugate of$a_1(f)$and $a_2(f)$respectively.

2. decompose the SDM at each frequency via Singular Value Decomposition (SVD). By taking the SVD of the spectral matrix, the spectral matrix is decomposed into a set of auto spectral density functions, each corresponding to a single degree of freedom (SDOF) system:

$$G(f) = USU^H = U \left[s_n^2 \right] U^H \tag{3}$$

where $U = [u_1, u_2, \ldots]$ represent the singular vectors (interpreted as the mode shapes) and **S** is a diagonal matrix containing the singular values s_n^2 (interpreted as the auto spectral densities of the modal coordinates).

The natural frequencies are estimated from peaks of the singular value plots which represent the auto PSD functions of equivalent single degree of freedom (SFOF) systems. For each identified natural frequency, a spectral bell centered on the peak is extracted from the corresponding singular value plot which is taken back in time domain by inverse Fast Fourier Transform (IFFT). This represents the auto-correlation function (ACF) of the corresponding equivalent SDOF system.

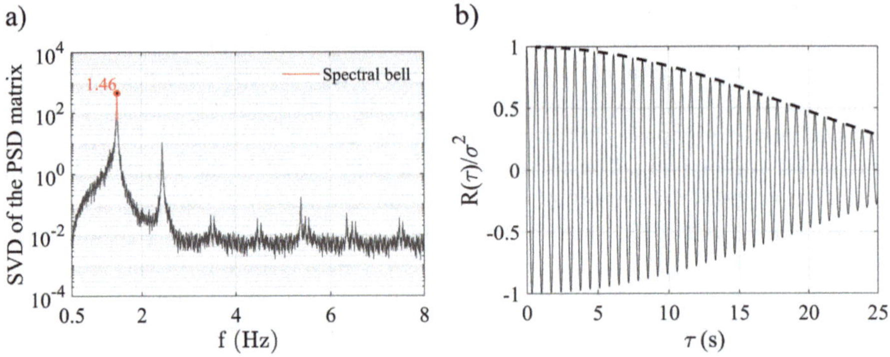

Fig. 7 December 23rd 2023: (**a**) Singular value plot (**b**) auto-correlation function

Table 3 Mean values and coefficient of variation of damping ratios

Damping ratio (%)	Mean value [%]	COV
ξ_x	0.56	0.08
ξ_y	0.53	0.14

3. estimate the damping ratio, δ which may be done by means of the logarithmic decrement technique where:

$$\delta = \frac{2}{k} \ln\left(\frac{r_0}{|r_k|}\right) \tag{4}$$

where r_0 is the initial value of the ACF and r_k is the k'th extreme.

Figure 7a shows the singular value plot corresponding to the December 23rd 2023 acceleration record (*y*-direction) depicted in Fig. 6a together with the spectral bell centered around the peak corresponding to the first flexural mode. The damping ratio is evaluated based on the auto-correlation function depicted in Fig. 7b of the equivalent SDOF system obtained by taking the spectral bell back to frequency domain by means of IFFT. The damping ratio corresponding to the *y* direction resulted $\xi_y = 0.51\%$ while for the *x* direction the damping ratio resulted $\xi_x = 0.49\%$.

Considering the ensemble of the 20 1-hr stationary acceleration records, the mean values and the coefficients of variation of the damping coefficients for both *x* and *y* directions are shown in Table 3. As may be seen, the results are in good agreement with the code-based recommendations [13].

Figure 8 shows the variation of the damping coefficients with mean wind speed (Fig. 8a) and root mean square (RMS) of acceleration (Fig. 8b) together with the corresponding linear trends (dotted lines). As indicated by Balio et al. [4] a slight dependence of the damping coefficient with wind velocity may be observed from Fig. 8a which may be attributed to the aerodynamic damping. This dependence may also be observed in relation to RMS of acceleration as depicted in Fig. 8b. The observed trends suggest that while structural damping dominates at lower

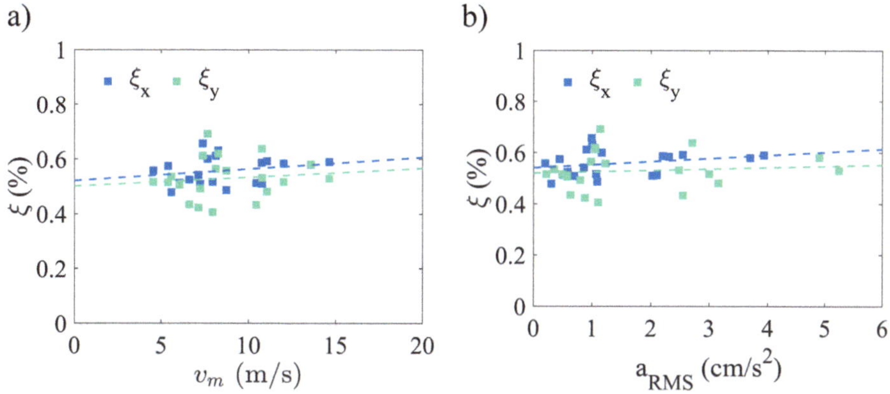

Fig. 8 Variation of the damping ratios with (**a**) mean wind speed (dotted lines show mean values of ξ) and (**b**) RMS of acceleration (dotted lines show linear trend of ξ)

wind speeds, aerodynamic effects become increasingly significant at higher wind velocities, consistent with previous findings for slender structures [4, 7]. However, the scatter in data points indicates that other factors, such as wind directionality, turbulence intensity, and the presence of ancillaries, may also influence the damping characteristics.

5 Conclusions

This paper presents the estimation of the dynamic characteristics of a triangular 50 m telecommunication lattice tower located in Afumați, Romania by means of the FDD technique applied to 20 stationary 1-hr acceleration records. The synchronous wind and acceleration measurements were acquired through a monitoring system installed on the tower comprising an ultrasonic anemometer and a triaxial accelerometer. On average, the damping coefficients resulted 0.56% in x-direction and 0.53% in y direction of the accelerometer which are aligned with one of the faces of the tower. The fundamental frequency of the tower estimated by making use of the Frequency Domain Decomposition technique resulted almost identical to that obtained by FEA. The results show a slight variation of the damping ratios with the mean wind speed and RMS of acceleration emphasizing the effect of aerodynamic damping for higher wind velocities. Finally, the overall damping ratios resulted similar to those reported in literature and are in good agreement with the values recommended by the Romanian wind code.

Acknowledgements This work was supported by a National Research Grants of the UTCB, project number $GnaC_{2023}^{ARUT} - UTCB - 06/2023$. The authors acknowledge the kind generosity of TECHNO VOLT company who donated the accelerometer used within this project.

References

1. Holmes, J.D.: Along-wind response of lattice towers—II. Aerodynamic damping and deflections. Eng. Struct. **18**(7), 483–488 (1996)
2. Shanmugasundaram, J., Harikrishna, P., Gomathinayagam, S., Lakshmanan, N.: Wind, terrain and structural damping characteristics under tropical cyclone conditions. Eng. Struct. **21**(11), 1006–1014 (1999)
3. Kareem, A., Gurley, K.: Damping in structures: its evaluation and treatment of uncertainty. J. Wind Eng. Ind. Aerodyn. **59**, 131–157 (1996)
4. Ballio, G., Maberini, F., Solari, G.: A 60 year old, 100 m high steel tower: limit states under wind actions. J. Wind Eng. Ind. Aerodyn. **43**, 2089 (1992)
5. Holmes, J.D., Schafer, B.L., Banks, R.W.: Wind-induced vibration of a large broadcasting tower. J. Wind Eng. Ind. Aerodyn. **43**, 2101–2109 (1992)
6. Glanville, M.J., Kwok, K.C.S.: Dynamic characteristics and wind induced response of a steel frame tower. J. Wind Eng. Ind. Aerodyn. **54–55**, 133–149 (1995)
7. Garcia, K.L., Maes, K., Parnas, V.E., Lombaert, G.: Operational modal analysis of a self-supporting antenna mast. J. Wind Eng. Ind. Aerodyn. **209**, 104490 (2021)
8. Tamura, Y., Kareem, A.: Advanced Structural Wind Engineering. Springer (2013)
9. Tamura, Y., Zhang, L., Yoshida, A., Nakata, S., Itoh, T.: Ambient vibration tests and modal identification of structures by FDD and 2DOF-RD technique. The Proceedings of the Symposium on the Motion and Vibration Control. **2003**, 8 (2002)
10. Brincker, R., Zhang, L., Andersen, P.: Modal identification from ambient responses using frequency domain decomposition. In: Proceedings of the International Modal Analysis Conference—IMAC, vol. 1, (2000)
11. Brincker, R., Ventura, C.E., Andersen, P.: Damping estimation by frequency domain decomposition. In: Proceedings of IMAC 19: A Conference on Structural Dynamics: February 5-8, 2001, Hyatt Orlando, Kissimmee, Florida, pp. 698–703 (2001)
12. Bendat, J., Piersol, A.: Random Data: Analysis and Measurement Procedures. John Wiley & Sons, New-York (1971)
13. Cod de proiectare. Evaluarea acțiunii vântului asupra construcțiilor, indicativ CR1-1-4/ 2012, Monitorul Oficial al României nr 704bis/15.X.2012
14. Calotescu, I., Bîtcă, D.: Characteristics of a thunderstorm downburst measured in a suburban area. Romanian J. Civil Eng. **16**(2), 174–183 (2025)
15. Solari, G., Piccardo, G.: Probabilistic 3-D turbulence modeling for gust buffeting of structures. Prob. Eng. Mech. **16**, 73–86 (2001)

Historical Evolution of Wind Design for Tall RC Chimneys in Romania

Florin Pavel

Abstract This study presents a historical overview on the wind design for the tallest structures built in Romania, namely the reinforced concrete chimneys used for both thermoelectrical powerplants and for industrial purposes, as well. The evolution of the design and construction practices of these structures are briefly presented and discussed. A historical overview of the procedures for wind design applied in Romania is briefly shown in the paper. Subsequently, a case-study involving two RC chimneys located in Bucharest and Iasi is presented. The wind force determined according to the various design regulations employed in Romania and the reliability to wind action are evaluated for two chimneys. The results show that the failure probability due to wind action as compared to the one associated to the seismic action is 10 larger for the chimney located in Iasi, while in the case of the chimney form Bucharest, the failure probabilities are almost identical ($6–8 \cdot 10^{-3}$).

Keywords Design codes · Reference wind pressure · Dynamic characteristics · Failure probability · Allowable stresses

1 Introduction

The reinforced concrete (RC) chimneys represent the tallest structures ever built in Romania. These structures were constructed for various industries, as well as for thermoelectric powerplants. The tallest RC chimney in Romania was built for industrial purposes in Baia-Mare, in the northwestern part of the country. In the case of RC chimneys for thermoelectric powerplants, the tallest structures had heights of 280 m and were built in various locations in Romania. The study of Wang and Fan [1] shows that most of the damage to reinforced concrete chimneys is due to

F. Pavel (✉)
Department of Reinforced Concrete Structures, Technical University of Civil Engineering Bucharest, Bucharest, Romania
e-mail: florin.pavel@utcb.ro

I. Calotescu, C. I. Coşoiu (eds.), *Proceedings of the Third National Conference on Wind Engineering*, Springer Proceedings in Earth and Environmental Sciences,
https://doi.org/10.1007/978-3-032-09671-5_7

temperature action, with wind action affecting more the steel chimneys. However, RC chimneys are also affected by wind action, and as the authors note the wind induced damage is significant for all types of chimneys. A significant part of the RC chimneys in Romania were built before 1977 when significant changes were made in the seismic design process. However, from the point of view of seismic damage, Schitco et al. [2] noticed a good overall behavior of RC chimneys during the Vrancea major earthquake of March 1977. The only damage observed on such a structure was a horizontal crack in the top part of a 120 m chimney located in Bucharest.

This paper shows a brief analysis of the evolution of the wind design in Romania with reference to RC chimneys. In addition, elements regarding the construction and the design itself of such structures are also presented. A case-study considering two chimneys located in Bucharest and Iasi is also shown in order to: (i) analyze the impact of the modifications of the wind action and (ii) evaluate the reliability to wind action.

2 Brief Presentation of Tall RC Chimneys Built in Romania

The first such structure made of cast-in-place concrete, namely a 70 m high chimney with a 3.6 m diameter was built in 1950 for a cement factory in Turda [3]. The shell thickness of this structure varied from 40 cm at the base to 15 cm at the top of the chimney. A brief history of the design and construction of the first RC chimneys is given in [4]. The technological evolution of the RC chimneys built in Romania is presented in a number of references in Romanian language (e.g [5–7].). The first cast-in-place RC chimneys built in the '50 s and '60 s were constructed using sliding formworks in a single tube variant which meant that the brick lining was adjacent to the reinforced concrete shell). The thermal insulation consisted of various powder materials. The tallest chimneys built in this variant had heights of 220 m [8]. After 1970, the double tube solution in which ventilated air passed between the brick lining and the RC shell was employed for the construction of all RC chimneys in Romania. The brick insulation was supported either by an independent frame structure or by structure supported by the RC shell which resembled a wheel with spokes (Fig. 1) [8].

Because of the necessity to expel larger and larger quantities of ash and other products and in order to ensure an adequate gas velocity a gradual increase in the height of the chimneys is observed. The last tall chimney was built in the mid '90 s for a thermoelectric powerplant located in Constanta. The height distribution of the RC chimneys is shown in Fig. 2 [8].

The location of the tallest RC chimneys in Romania for both thermoelectrical powerplants and industrial purposes is illustrated in Fig. 3. A concentration of tall RC chimneys is in the south-western part of Romania where the largest thermoelectric powerplants were built in the vicinity of the coal deposits located in that region. An additional observation made in the paper of Agent and Rosetnic [8] was that the taller chimneys were generally located in areas of low and moderate seismicity.

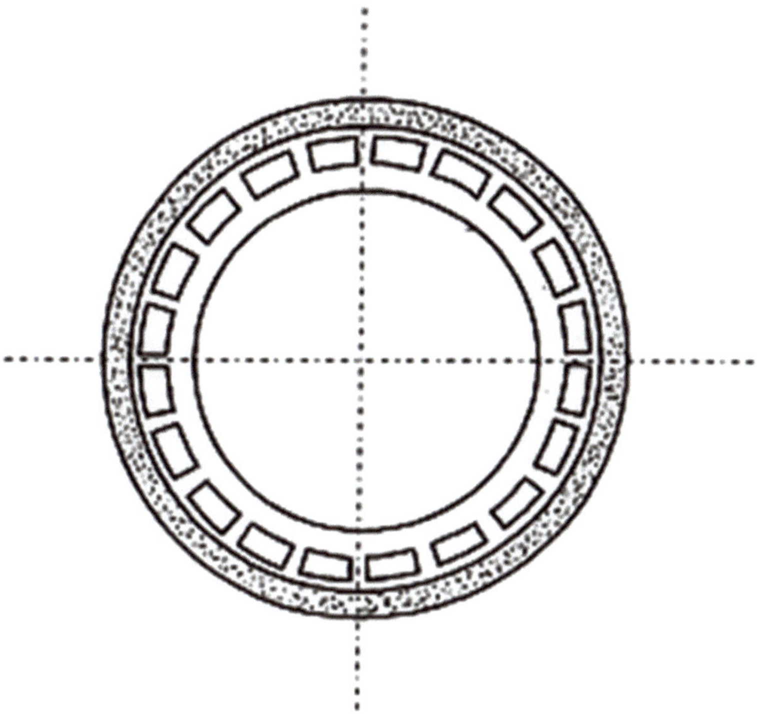

Fig. 1 Cross-section of RC chimneys built in the ‘80 s in Romania [8]

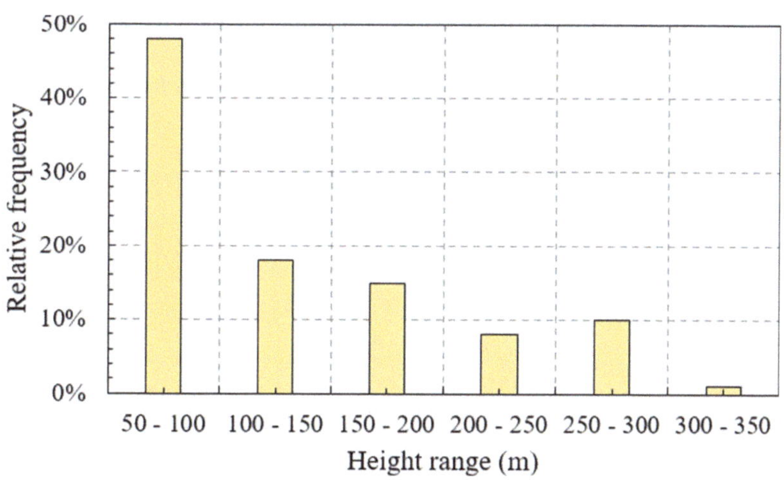

Fig. 2 Height distribution of RC chimneys in Romania [8]

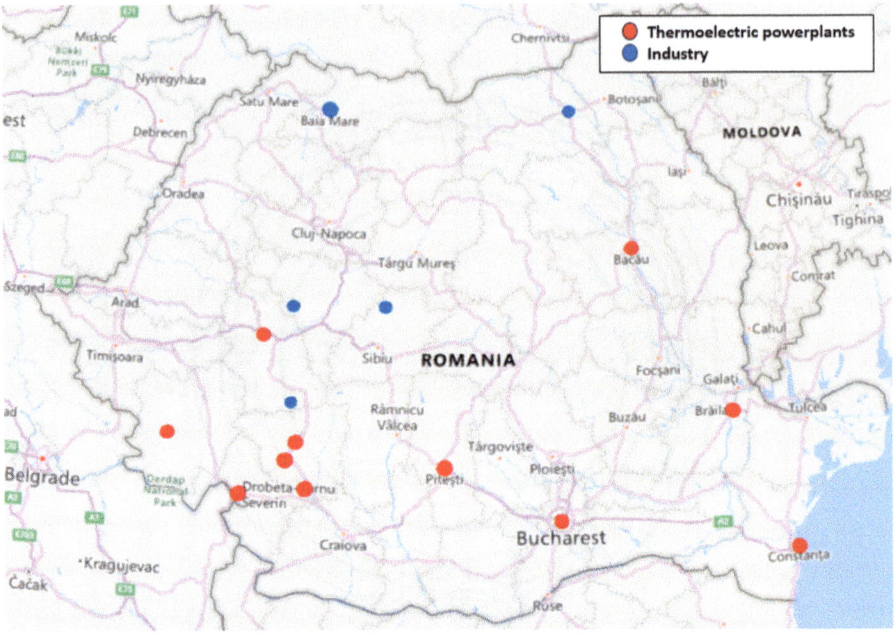

Fig. 3 Location of the tallest RC chimneys in Romania

3 Brief Evolution of Wind Action Design in Romania

The wind action was considered for the design of structures in Romania well before the enforcement of the first national code, namely STAS 946, enforced for the first time in 1950 [9] and subsequently modified in 1956 [10]. Thus, the Technical Indicator for Constructions [11] assigns the following relation for the evaluation of the wind pressure p_w (in kg/m^2) for chimneys:

$$p_w = 120 + 0.6 \cdot h \tag{1}$$

where h is the height of the chimney in meters. The wind force F_w acting on a chimney was computed as:

$$F_w = n \cdot A_w \cdot p_w \tag{2}$$

where A_w is the area exposed to wind action and n is taken as 0.67 for circular chimneys, 0.71 for octagonal chimneys and 1.0 for square or rectangular sections. The above-mentioned relations were employed in the design of the first tall chimney made of cast-in-place concrete constructed in Romania in 1950 [12].

The first version of the Romanian wind design code, namely STAS 946-50 [9] included provisions regarding the evaluation of the accidental loads (wind, snow and temperature). The name of the code was changed in the 1956 revision because for some structures, like the chimneys, the wind load was a fundamental load and not accidental. The design wind pressure from STAS 946-50 [9] was computed with the following relation:

$$p_w = g_v \cdot k \cdot k_1 \cdot k_2 \tag{3}$$

where g_v was the reference dynamic wind pressure (either 0.6 or 0.8 kN/m^2 at 20 m above ground level), k was a pressure coefficient depending on the shape of the construction, k_1 was a coefficient that took into account the position of the construction with respect to the ground and k_2 was a dynamic coefficient depending on the fundamental eigenmode of the construction (for chimneys $k_2 = 1.5$). For slender structures, it was recommended that the fundamental eigenperiod should not be situated in the interval b/g_v and $1.5b/g_v$, where b was the section diameter.

STAS 945-56 [10] proposed the same relation for the evaluation of the design wind pressure as the previous 1950 version of the code. The reference dynamic wind pressure g_v had two values either 0.5 or 0.7 kN/m^2 at 10 m above ground level. In addition, the values of the pressure coefficient k were modified for some specific shapes. A critic valid for both STAS 946-50 [9] and STAS 946-56 [10] mentioned in the study [4] was that even though a gust coefficient had to be computed for the evaluation of the wind action, no guidance on how to evaluate the modal characteristics of the chimneys were given. The design wind action was decomposed into three components (an axial symmetric component, an axial antisymmetric component and a period component) each generating a particular type of stress. The decomposition of the wind action is illustrated in Fig. 4 [13]. In the subsequent modification of the code from 1975 and 1978 [14], a new zonation map of the reference wind pressure (with five zones instead of two) was proposed.

The design wind pressure was evaluated using the relation [14]:

$$p_w = g_v \cdot \beta \cdot c_n \tag{4}$$

where β was a dynamic coefficient and c_n was a coefficient of aerodynamic force acting on surface. The dynamic coefficient for tall and slender structures was computed as [14]:

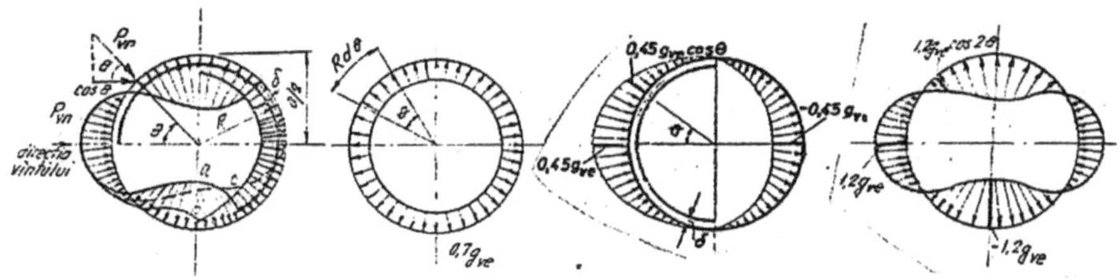

Fig. 4 Decomposition of wind action [13]

$$\beta = 1 + \xi \cdot r \tag{5}$$

In Eq. 5, ζ was an amplification coefficient (depending on the fundamental eigenperiod of the structure and on the structural system) and r is the gust factor.

The reference wind pressure was given separately for urban and non-urban areas. In the case of urban areas, the range of reference wind pressures was between 0.34 and 0.61 kN/m^2, while in the case of non-urban areas the range was 0.45–0.85 kN/m^2. Also, in the 1975 revision of the wind design code, the across-wind load had to be evaluated for tall and slender structures. The critical wind velocity v_{cr} corresponding to the resonance was evaluated as [14]:

$$v_{cr} = \frac{d}{St \cdot T_1} \tag{6}$$

where d was the diameter of the construction, St is the Strouhal's number and T_1 is the fundamental eigenperiod. The resonance check had to be performed if the critical velocity was less than 26 m/s. The across-wind pressure was then evaluated as [14]:

$$p_t = 0.16 \cdot \beta'^2 \cdot p_{cr} \cdot d \cdot \frac{z}{h} \tag{7}$$

where p_{cr} was the critical wind pressure (computed based on the critical wind velocity) and z is the height at which the evaluation is made. β' is a dynamic amplification coefficient computed as [14]:

$$\beta' = \frac{3.14}{\delta_I} \tag{8}$$

where δ_I was the logarithmic decrement of damping for the oscillations corresponding to the fundamental eigen mode. The values of δ_I ranged between 0.1 for steel structures to 0.4 for masonry structures.

The along-wind design wind pressure from the next code version, STAS 10101-20/90 [15] was determined with the following relation:

$$p_w = g_v \cdot \beta \cdot c_{ni} \cdot c_h(z) \tag{9}$$

where β was a gust coefficient, c_{ni} was an aerodynamic coefficient acting on surface i and $c_h(z)$ was a coefficient of variation of the dynamic wind pressure with the height. The reference dynamic wind pressure was between 0.3–0.55 kN/m^2 and was evaluated based on a reference wind velocity with a mean recurrence interval of 10 years. The across-wind check was performed using an updated relation as compared to the previous code version [15]:

$$p_t = 0.25 \cdot p_{cr} \cdot d \tag{10}$$

Table 1 Evolution of the reference wind pressure in kN/m^2 for 10 large cities in Romania

City	STAS 946-50 [9]	STAS 946-56 [10]	STAS 10101-20/78 [14]	STAS 10101-20/90 [15]	NP082-04 [16]	CR 1-1—4/2012 [17]
Brasov	0.6	0.5	0.34	0.3	0.4	0.6
Bucharest	0.6	0.7	0.41	0.42	0.5	0.5
Cluj-Napoca	0.6	0.5	0.34	0.3	0.4	0.5
Craiova	0.6	0.7	0.41	0.42	0.5	0.5
Constanta	0.8	0.5	0.53	0.55	0.5	0.5
Galati	0.6	0.7	0.41	0.42	0.5	0.6
Iasi	0.6	0.5	0.41	0.55	0.7	0.7
Oradea	0.6	0.5	0.34	0.3	0.5	0.5
Ploiesti	0.6	0,7	0.34	0.42	0.4	0.4
Timisoara	0.6	0.5	0.41	0.3	0.4	0.6

NP082-04 [16] increased the mean recurrence interval of the reference wind velocity to 50 years, an interval which is also used by the current version of the wind design code in Romania. The current version of the Romanian wind design code, CR 1-1-4/2012 [17] has the same format as the Eurocode 1-1-4 [18]. A comparison of the zonation maps of the various wind design codes enforced in Romania in terms of reference wind pressure can be found in the study of Aldea et al. [19].

In this context, a comparison of the evolution of the reference wind pressures for 10 large cities in Romania is given in Table 1. From the data presented in Table 1, it can be observed that: the range of reference wind pressures is between 0.3–0.7 kPa and that significant changes have occurred from one version of the code to the other (both increases and decreases of the reference wind pressure have occurred). It has to be mentioned, as well, that other tall and flexible structures, like the steel towers for electricity lines were designed using other regulations proposed by the electricity operator. For some areas, the design wind velocity (which was a function of the powerline tension) in these regulations was significantly larger than the one given in the design codes used for constructions [7].

In the paper of Aldea et al. [19] are mentioned also the maximum wind velocities for some meteorological stations in Romania located in areas with altitudes less than 300 m. The maximum wind velocities are of the order of 120 km/h (about 35 m/s), which corresponds to a wind pressure of about 0.7 kPa. An area in which large wind velocities have been observed is the south-western part of Romania (including Timisoara), where values of the order of 145 km/h and even 165 km/h have been recorded, and significant damage was observed for the electricity network. The last wind velocity was recorded during a storm in June 1982 at the meteorological station located at Timisoara Airport which was able to function for wind velocities in excess of 40 m/s [20]. For comparison purposes, the most recent significant storm from the same part of Romania (from September 2017) generated peak wind velocities of the order of 120 km/h [21].

4 Evolution of Design Procedures for RC Chimneys in Romania

The last versions of the dedicated design code for industrial RC chimneys P133 were enforced in 1997 [22] and finally in 2005 [23]. Before these versions of the design codes, STAS 3466-58 [24] and STAS 3466-68 [25] based on the allowable stress method (even though at that time the ultimate strength method and the limit states method were employed for RC structures) were used as reference. Some guidance on the design of RC chimneys were also given in an older reference [26] published before the enforcement of the design codes. An important point which has to be mentioned is that the design of the first RC chimneys in Romania was done using a load combination in which both the wind action and the seismic action were considered as acting simultaneously. According to STAS 3466-58 [23], the following load combinations had to be considered for design:

- Load combination A: self-weight + wind;
- Load combination B: self-weight + wind + temperature;
- Load combination C: self-weight + earthquake + temperature.

The allowable stresses in the concrete and steel corresponding to each of the three above-mentioned load combinations from STAS 3466-58 [24] are given in Table 2 as a function of the characteristic concrete f_{ck} and steel strengths f_{yk}.

Before the enforcement of STAS 3466-58 [24], it is mentioned by Hangan [27] that the load combination C included also 50% of the wind load besides the earthquake and temperature action. The same load combinations from STAS 3466-58 [24] were also employed in the subsequent version of the code (STAS 3466-68 [25]).

From the point of view of the dynamic characteristics of the Romanian RC chimneys, Rosetnic [28] proposed the following empirical relation for the evaluation of the fundamental eigenperiod:

$$T_1 = -0.354 + 0.019 \cdot h \tag{11}$$

Vlad and Vlad [29] developed another relation for the computation of the fundamental eigenperiod of Romanian RC chimneys:

$$T_1 = 1.7 \cdot h/100 \tag{12}$$

Table 2 Allowable stresses in concrete and steel for the three load combinations according to STAS 3466-58 [24]

Load combination	Allowable stresses	
	Concrete	Reinforcing steel
A	$0.3f_{ck}$	$0.5f_{yk}$
B—horizontal sections	$0.4f_{ck}$	$0.65f_{yk}$
B—vertical sections	$0.25f_{ck}$	$0.5f_{yk}$
C	$0.45f_{ck}$	$0.7f_{yk}$

ACI 307-08 [30] employs for the evaluation of the fundamental eigenperiod the following equation which depends on more input parameters:

$$T_1 = 5\frac{h^2}{\overline{d}(b)} \cdot \sqrt{\frac{\rho_{ck}}{E_{ck}}} \cdot \left[\frac{t(h)}{t(b)}\right]^{0.3} \quad (13)$$

where h is the height of the chimney above base, $t(h)$ and $t(b)$ are the thicknesses at the top and bottom part of the chimney, $\overline{d}(b)$is the mean diameter at the bottom part, ρ_{ck} is the density of the concrete and E_{ck} is the modulus of elasticity of the concrete. The three above-mentioned relations provide rather similar results for chimneys up to a height of 250 m.

5 Case-study

An example for the evaluation of the wind force according to the various wind design codes enforced in Romania is shown using as case-study two RC chimneys used for thermoelectric powerplants in Bucharest and Iasi. Both chimneys have also been employed in the study of Pavel et al. [31] in which the assessment of the seismic reliability of the two constructions is performed. In addition, a simple reliability assessment for the wind action is performed for both structures. The chimney located in Bucharest was built in 1972, has a height of 180 m and has an exterior diameter ranging from 12.7 m at the base to 8.5 m at its top. The thickness of the shell element is in the range 25–60 cm. The chimney located in Iasi was built in 1984, has a height of 164 m and has an exterior diameter ranging from 14.1 m at the base to 11.8 m at its top. The thickness of the shell element is in the range 26–58 cm.

The pressure distribution along the height for the two chimneys using the design codes enforced in Romania is illustrated in Fig. 5. Significant differences in terms of the shape of the wind pressure distribution with the height and the maximum wind pressure values along the height can be observed from Fig. 5. The along-wind force associated to the two chimneys and computed using the various design codes employed in Romania is shown in Fig. 6. It can be observed from Fig. 6, the range of along-wind forces varies significantly.

A simple reliability assessment of the analyzed RC chimneys to wind action is performed considering the wind data for two weather stations located in Bucharest and Iasi and which are given in the study of Aldea et al. [19]. The mean values and the coefficients of variation for the wind velocity (computed based on about 40 years of observations) for Bucharest and Iasi are taken from Aldea et al. [19] and are reported in Table 3 together with the reference wind velocity from the code CR 1-1-4/2012 [17].

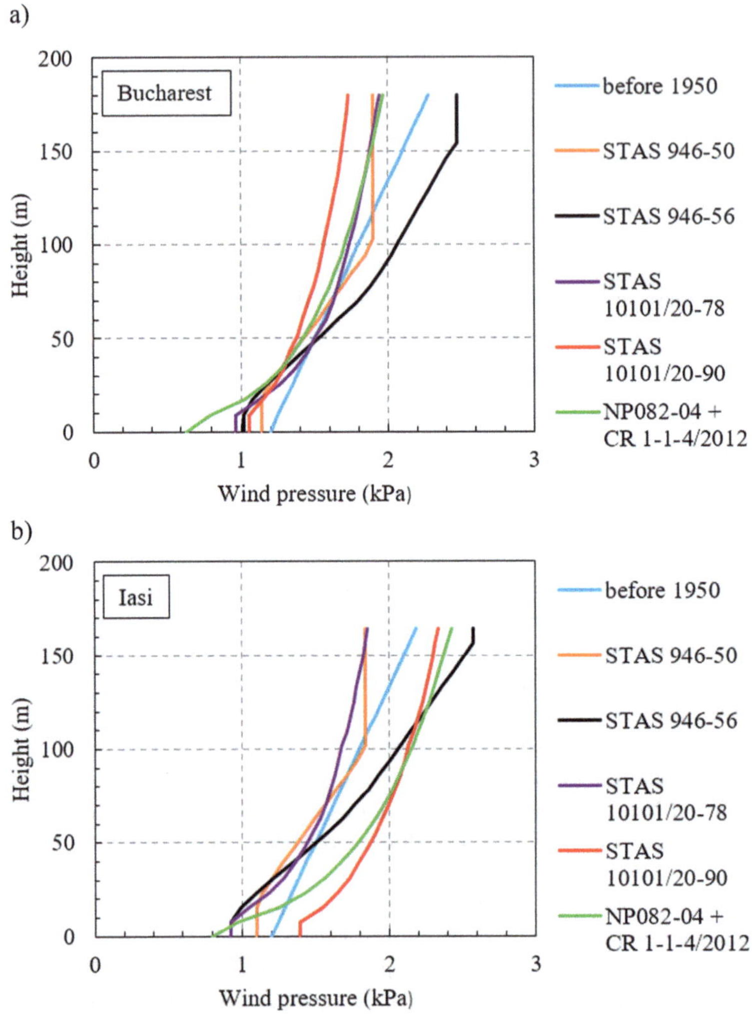

Fig. 5 Comparison of the pressure distribution along the height evaluated using the various design codes enforced in Romania: (**a**) Bucharest chimney; (**b**) Iasi chimney

The reliability to wind action was evaluated using the following approach:

- the wind velocity data for each site was based on sampling (the sampling was performed considering the statistical indicators from Table 2). A lognormal probability distribution was considered for sampling the wind velocity [32];
- the bending moments were computed from the wind forces (along-wind and cross-wind) evaluated using the approach from CR 1-1-4/2012 [17];
- the bending moment capacities of the RC section (evaluated considering mean material strengths at the bottom part of the chimney) were compared against the bending moments due to wind action.

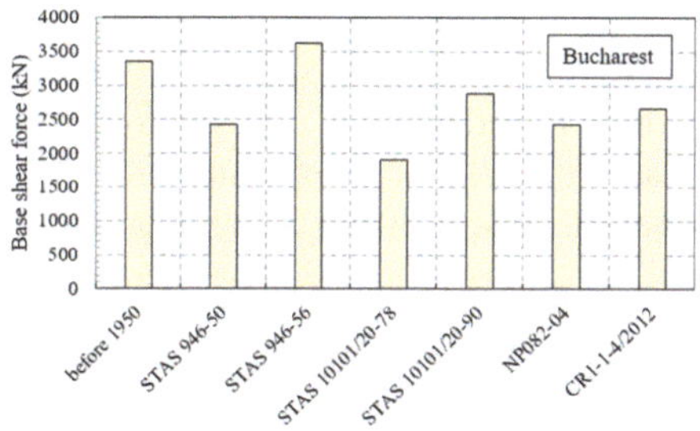

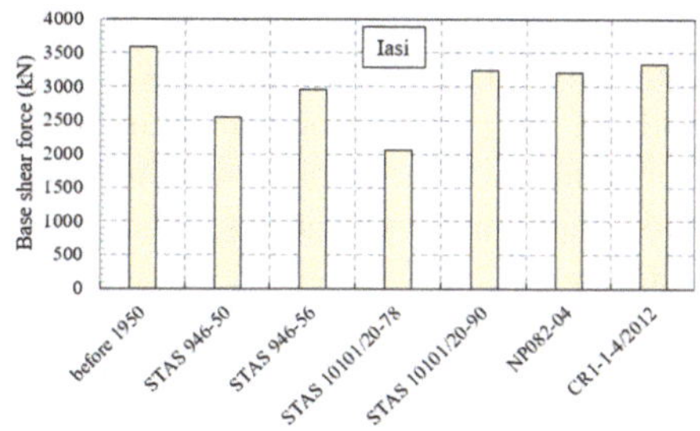

Fig. 6 Comparison of the along-wind force evaluated using the various design codes enforced in Romania: (**a**) Bucharest chimney and (**b**) Iasi chimney

Table 3 Wind data for Bucharest and Iasi

Location of chimney	Mean wind velocity (m/s) [19]	Coefficient of variation of wind velocity [19]	Reference wind velocity (m/s) [17]
Bucharest	14.0	0.26	28.3
Iasi	16.6	0.31	33.5

Table 4 Annual failure probabilities due to wind and earthquake action for the two chimneys located in Bucharest and Iasi

Location of chimney	Annual failure probability—yielding due to wind action	Annual failure probability—yielding due to earthquake action [31]
Bucharest	$6.10 \cdot 10^{-3}$	$7.85 \cdot 10^{-3}$
Iasi	$9.00 \cdot 10^{-3}$	$8.55 \cdot 10^{-4}$

The annual probabilities of exceedance of yielding for the two RC chimneys are given in Table 4. The probabilities were computed using a simple R-S reliability problem [33] and considering a lognormal probability distribution for the bending moment capacity of the section. In Table 4 are also given the annual exceedance probabilities associated to yielding due to seismic action taken from the study of Pavel et al. [31]. From Table 4 it can be observed that the annual failure probability due to wind action has the same order of magnitude for both chimneys. In addition, the failure probability due to wind action as compared to the one associated to the seismic action is significantly larger for the chimney located in Iasi. In the case of the Bucharest chimney, the two failure probabilities are almost identical.

6 Conclusions

This study presents a brief overview on the wind design of the tallest structures ever built in Romania, namely the RC chimneys employed by various industries, as well as by thermoelectric powerplants. In addition, the construction practice and the design procedures employed in Romania are discussed, as well. A case-study considering two RC chimneys located in Bucharest and Iasi is shown in order to evaluate the wind action according to the various generations of codes enforced in Romania and to determine in a simplified manner the reliability to wind action. The most important observations and conclusions of this research can be summarized as follows:

- Wind action was considered in the design of RC chimneys even before the enforcement of the first wind design code STAS 946–50 [9];
- A net evolution in terms of construction practice and technology can be noticed for the period 1960–1995;
- Significant wind velocities in excess of 45 m/s have been recorded at meteorological stations located in the south-western part of Romania, an area which is characterized by large design wind pressures;
- Two simple empirical relations which depend only on the height of the RC chimney can be used for the estimation of the fundamental eigenperiods;
- The range of along-wind forces computed for the two chimneys based on the various versions of the Romanian wind design code varies significantly. In addition, significant differences in terms of the shape of the distribution of the wind pressure with the height and the maximum wind pressure values along the height have also been observed;
- The reliability assessment of two RC chimneys shows that the RC chimney located in Bucharest has similar annual failure probabilities for yielding due to wind and seismic action. In addition, the failure probability due to wind action of the chimney located in Bucharest is smaller than that of the chimney located at Iasi.

Acknowledgements The support of Prof. Valeriu Rosetnic from the Technical University of Civil Engineering Bucharest, who kindly provided valuable information regarding the design of RC chimneys is gratefully acknowledged.

References

1. Wang, L., Fan, X.-Y.: Failure of high chimneys: a review. Eng. Fail. Anal. **105**, 1107–1117 (2019)
2. Schitco, V., Ghergu, I., Badescu, F.: The seismic behaviour of constructions for thermoelectric powerplants. In: Proceedings of the National Conference for Design, Execution and Expertise of Structures in Seismic Areas, Iasi, Romania, vol. 1, pp. 301–310 (1983)
3. Avram, C.: Reinforced concrete in Romania. Tehnica, Bucharest, Romania (1987)
4. Hangan, S.: Aspects from the execution of precast chimneys in RPR. Constructions. **9**(5), 258–262 (1958)
5. Naum, L.: New orientations in selecting the constructive solutions of industrial chimneys. Constructii. **1**, 5–11 (1981)
6. Venat, A.: New technological solutions for constructing industrial reinforced concrete chimneys by sliding. Constructions. **8**, 5–9 (1983)

7. Ganea, I., Croitoru, M., Savin, F.: The Electrification Process in Romania 1951–1992. Tehnica, Bucharest, Romania (1996)
8. Agent, R., Rosetnic, V.: Problems of the technical expertise of RC chimneys for thermoelectric powerplants. In: Proceedings of the National Symposium for Seismic Risk Reduction of Industrial Constructions, pp. 71–84. Bucharest, Romania (1998)
9. OSS.: STAS 946–50: Loads in constructions. Accidental loads. Bucharest, Romania (1950)
10. OSS.: STAS 946–56: Loads in constructions. Climatic loads. Bucharest, Romania (1956)
11. Asquini, V.: Technical indicator for constructions. Cartea Romaneasca, Bucharest, Romania (1938)
12. Agent, R., Popovici, G.: The construction of a 70 m high factory RC chimney. Constructii. **2**(8), 5–16 (1951)
13. Borza, N.: Special reinforced concrete structures—lecture notes. Tall RC chimneys. ICB, Bucharest, Romania (1972)
14. IRS.: STAS 10101/20-78: Actions on constructions. Wind action. Bucharest, Romania (1978)
15. IRS.: STAS 10101/20-90: Actions on constructions. Wind action. Bucharest, Romania (1990)
16. MTCT.: NP-082-04. Design code. Basis of design and actions on constructions. Wind action. Bucharest, Romania (2005)
17. MDRT.: CR 1-1-4/2012: Design code for the evaluation of wind action on buildings. Bucharest, Romania (2012)
18. CEN.: Eurocode 1: Actions on structures. Part 1–4: General actions - wind actions. Brussels, Belgium (2005)
19. Aldea, A., Demetriu, S., Văcăreanu, R., Lungu, D., Arion, C., Neagu, C.: A Short Historical Overview of Wind Characteristics and Wind Pressure for Design in Romania. Tech. Trans. **116**, 97–115 (2019)
20. Putnoky, E.: Events produced at the IRE Timisoarea installations as a result of the June 27, 1982 storm. Energetica. **31**(1), 20–26 (1983)
21. Balazs, R.A.: Storm phenomenon in the Western Plain. Rev. Stiint. ANM 41–60. ISSN 2069-962X (2018)
22. MLPAT.: P133-96: Code for the design of industrial cast in-place RC chimneys. Bucharest, Romania (1997)
23. MTCT.: P133-04: Code for the design of industrial cast in-place RC chimneys. Bucharest, Romania (2005)
24. OSS.: STAS 3466-58: Independent chimneys. Design prescriptions. Bucharest, Romania (1958)
25. OSS.: (1968) STAS 3466-68: Independent chimneys. Design prescriptions. Bucharest, Romania (1968)
26. Ganea, N.: Concrete. In: Gorjanu, I.P. (ed.) Design and Execution of Concrete and Reinforced Concrete Constructions, Bucharest, Romania, vol. 2, (1948)
27. Hangan, M.: Reinforced concrete. Industrial constructions. Tehnica, Bucharest, Romania (1958)
28. Rosetnic, V.: An assessment of EC8 rules for the seismic design of reinforced concrete chimneys. Bull. Tech. Univ. Civ. Eng. Buchar. **1**, 52–58 (2001)
29. Vlad, I., Vlad, M.: In situ measurements for the structural assessment of existing tall industrial Chimneys. In: Proceedings of the 15th World Conference on Earthquake Engineering. Lisbon, Portugal (2012)
30. ACI Committee 307.: ACI 307–08: Code Requirements for Reinforced Concrete Chimnseys (ACI 307–08) and Commentary. Farmington Hills, USA (2008)
31. Pavel, F., Nica, G., Noroozinejad Farsangi, E.: Seismic reliability assessment of existing RC chimneys in Romania. J. Build. Pathol. Rehabil. **8**, 88 (2023)
32. JCSS.: JCSS Probabilistic model code. Joint Committee on Structural Safety, ISBN 978-3-909386-79-6 (2001). https://www.jcss-lc.org/jcss-probabilistic-model-code/
33. Melchers, R., Beck, A.: Structural reliability analysis and prediction. 3rd John Wiley & Sons (2018)

Assessment of Wind Speed Between Satellite Altimetry Data and SWAN Model Simulations in the Black Sea Basin

Eduard Luca, Georgeta Bandoc, Mircea Degeratu, and Iris Ţiţineanu

Abstract The objective of this study is to assess wind speed patterns within the Black Sea region spanning from 2019 to 2021. To fulfil this objective, satellite altimetry data on 10-meter wind speed (U_{10}) were sourced from the IMOS database and used together with empirical estimation of wind speed derived from the outputs of significant wave height (H_s) values based on SWAN model simulations. The findings resulted from the comparative analysis of these datasets enhance our comprehension of wind dynamics within the Black Sea domain. Such insights hold considerable scholarly, economic, geopolitical, and strategic implications, particularly in light of the geostrategic significance of this geographical area.

Keywords Wind speed · Numerical modelling · Black Sea · Satellite altimetry

E. Luca (✉)
University of Bucharest, Faculty of Geography, Doctoral School "Simion Mehedinţi", Center for Coastal Research and Environmental Protection, Bucharest, Romania

G. Bandoc
University of Bucharest, Faculty of Geography, Doctoral School "Simion Mehedinţi", Center for Coastal Research and Environmental Protection, Bucharest, Romania

Romanian Association for Wind Engineering, Bucharest, Romania

Academy of Romanian Scientists, Bucharest, Romania

M. Degeratu
Academy of Romanian Scientists, Bucharest, Romania

Technical University of Civil Engineering Bucharest, Bucharest, Romania

I. Ţiţineanu
University of Bucharest, Faculty of Geography, Doctoral School "Simion Mehedinţi", Center for Coastal Research and Environmental Protection, Bucharest, Romania

Meteo Romania (National Meteorological Administration), Bucharest, Romania

I. Calotescu, C. I. Coşoiu (eds.), *Proceedings of the Third National Conference on Wind Engineering*, Springer Proceedings in Earth and Environmental Sciences,
https://doi.org/10.1007/978-3-032-09671-5_8

1 Introduction

Satellite data, integral to both operational tasks and scientific inquiry, undergoes meticulous processing stages, encompassing acquisition, homogenization, quality control, cross-calibration, product generation, and final quality assurance. This comprehensive procedure involves the application of diverse algorithms to yield a refined gridded product.

Within wind wave modelling, third-generation models like SWAN, WAM, and WAVEWATCH III are pivotal for examining wave forecasts and climate phenomena [1, 2]. The SWAN (Simulating WAves Nearshore) model is designed for coastal and nearshore applications and incorporates shallow-water physics such as wave breaking and bottom friction [3]. The WAM (Wave Model) is the first fully third-generation wave model that simulates the evolution of the wave spectrum based on the energy balance equation and nonlinear wave-wave interactions [4]. The WAVEWATCH III model is a spectral wave model developed by NOAA for operational wave forecasting and is capable of simulating wave generation and propagation over global ocean scales and is known for its modular structure and flexibility in incorporating advanced physics Schemes [5]. Notably, the significant wave height (H_s) stands as a prominent parameter in addition to others such as wave period (T) or wave direction that further characterize wave dynamics. Prior studies extensively deliberated on wind wave simulation strategies in coastal and open sea domains [6]. Comparative analyses of numerical wave models, satellite data, and in-situ measurements revealed substantial correlations, underscoring the reliability of these datasets. A study by from 2015 [7] scrutinized storm cases in the Western

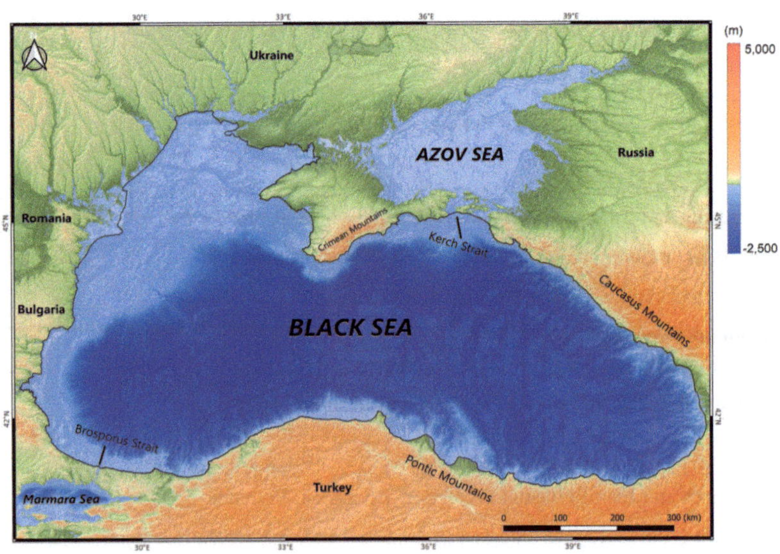

Fig. 1 Black Sea topographic map

Black Sea, contrasting SWAN numerical wind wave simulations with satellite data and in-situ measurements. In Fig. 1 the geographical area of Black Sea is illustrated.

2 Satellite Altimetry

This study utilizes 10-meter wind speed (U_{10}) data from satellite altimetry, which is a technique measuring the time taken for radar signals to travel from the satellite antenna to the Earth's surface and back. This method yields a wealth of information crucial for diverse applications, particularly in assessing oceanic and sea conditions, as explained in Lee et al. [8]. Data was sourced from the IMOS (Integrated Marine Observation System) database, maintained by the Department of Infrastructure Engineering at the University of Melbourne, Australia [9].

This database, established in 2019 and regularly updated, offers global datasets on wind speed (U_{10}) and significant wave height (H_s) from various altimetry missions launched since 1985. The data contains measurements in the Ku-band and C-band, facilitating the determination of parameters such as wave periods and wind speed near the sea surface. For this study, wind speed data from four satellites covering the Black Sea basin between January 1, 2019 and December 31, 2021, were processed, details in Table 1.

The satellite ground tracks over the geographical area of the Black Sea for the studied time interval are exemplified in Fig. 2 which includes only the Jason-3 and Sentinel-3A missions. The inclusion of all four missions and the full temporal extent would have resulted in an overly dense map, making the visualization of individual satellite tracks difficult. The use of data from satellite altimetry from the IMOS database has been conducted in other studies such as [10, 11].

3 Numerical Wind Wave Modelling

Numerical wave models serve as indispensable tools for understanding wind wave phenomena, encapsulating underlying physical principles. Originating in 1947 with Sverdrup and Munk's [12] statistical technique focusing solely on wind wave height,

Table 1 List of altimetry satellites

Satellite	Launch date (dd/mm/yyyy)	End date (dd/mm/yyyy)	No. of processed measurements
SARAL	25/02/2013	Ongoing mission	64851
Jason-3	17/01/2016	Ongoing mission	70442
Sentinel-3A	16/02/2016	Ongoing mission	68756
Sentinel-3B	25/04/2018	Ongoing mission	67898

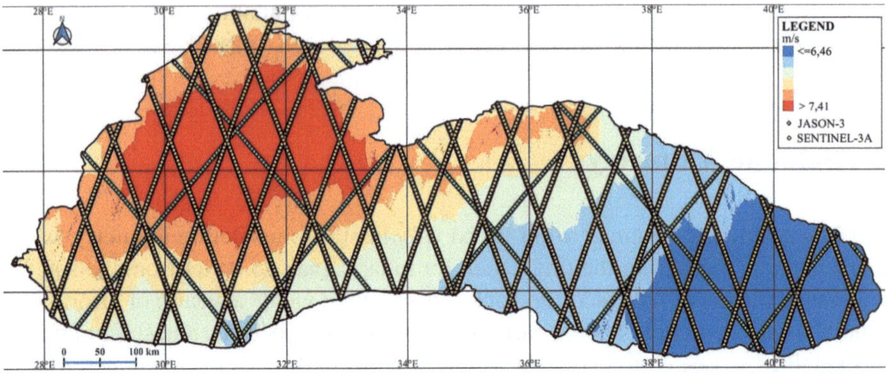

Fig. 2 Density of altimetry points from Jason-3 and Sentinel-3A over Black Sea basin for the year 2020

early models utilized coarse grids, favoring deep-water regions. However, precise modelling, particularly in coastal zones, demands finer resolution, necessitating complex computations.

In this study, the SWAN (Simulating WAves Nearshore) model was used, a third-generation numerical model designed for realistic wave parameter estimation in coastal zones, lakes, and estuaries. Developed at the Delft University of Technology in the Netherlands, SWAN operates by integrating the energy conservation equation, assuming the absence of currents. The model can simulate wave generation, propagation, and transformation by accounting for factors such as triads and quadruplet wave-wave interactions, wave-current interactions, variable bathymetry, whitecapping, bottom friction and other physical processes [13].

The SWAN model (version 41.31AB) with a six-hour time step and on a structured mesh with a spatial resolution of 0.05×0.05 was applied for the Black Sea basin using input data on wind and waves sourced from the ERA5 reanalysis of ECMWF (European Centre for Medium-Range Weather Forecasts) and bathymetric data from GEBCO (General Bathymetric Chart of the Oceans). A more detailed description of the model setup, including the spatial domain, resolution, bathymetric characteristics, and physical processes and their parameters, is provided in another study [10]. For this current paper, the same configuration was adopted to ensure consistency and comparability.

The main physical processes activated in the model include whitecapping dissipation, depth-induced wave breaking, shallow water dissipation, and nonlinear wave-wave interactions, along with adjustments of their tunable parameters.

For clarity, the key coefficients are as follows: *cds2* and *delta* control the whitecapping dissipation rate. *Cn4* and *iquad* refer to the quadruplet wave–wave nonlinear interaction configuration, a*lpha* and *gamma* are used in the depth-induced breaking formulation, bottom friction is governed by a constant friction coefficient, *trfac* and *cutfr* are used in the triad interaction scheme, where *trfac* is the triad scaling factor and *cutfr* defines the cut-off frequency below which interactions are considered [13].

The SWAN model solves the spectral action balance equation which describes the conservation of wave action density [14, 15]:

$$\frac{\partial}{\partial t}N + \frac{\partial}{\partial x}c_x N + \frac{\partial}{\partial y}c_y N + \frac{\partial}{\partial \sigma}c_\sigma N + \frac{\partial}{\partial \theta}c_\theta N = \frac{S}{\sigma} \tag{1}$$

where the term $\frac{\partial}{\partial t}N$ describes the local rate of change of the action density in time. The second and the third terms ($\frac{\partial}{\partial x}c_x N$ and $\frac{\partial}{\partial y}c_y N$) represent the propagation of wave action in geographical space, with c_x and c_y as the propagation velocities in the x and y directions, respectively. The term $\frac{\partial}{\partial \sigma}c_\sigma N$ accounts for shifting of the relative frequency due to variations in depth and currents; here, c_σ is the propagation speed in the frequency space. The final term on the left side $\frac{\partial}{\partial \theta}c_\theta N$ describes depth and current-induced refraction, with c_θ being the propagation speed in the directional space. The right-side term $\frac{S}{\sigma}$ represents the effects of generation, dissipation, and nonlinear wave-wave interactions.

The six source terms representing deep and shallow water physical processes, which allow the simulation of wind-to-wave energy transfer and various wave energy dissipation mechanisms (S_{tot}), are expressed as follows:

$$S_{tot} = S_{in} + S_{wc} + S_{nl4} + S_{bot} + S_{brk} + S_{nl3} \tag{2}$$

Here, the deep-water source terms (S_{in}, S_{wc}, and S_{nl4}) correspond respectively to the energy input from wind to waves, energy dissipation due to whitecapping, and non-linear quadruplet wave–wave interactions. The shallow water source terms (S_{bot}, S_{brk}, and S_{nl3}) represent energy dissipation caused by bottom friction, depth-induced wave breaking, and non-linear triad interactions, respectively. Further details on the formulation of these processes can be found in other studies [1, 16].

The choice of SWAN model over other third-generation models such as WAM or WAVEWATCH III is justified by its ability to simulate nearshore and coastal wave processes in semi-enclosed basins like the Black Sea, where depth-induced effects and shallow water dissipation play a significant role. While WAM and WW3 are widely used for deep ocean and large-scale applications [4, 17, 18], SWAN has been shown to perform more accurately at a regional scale, enclosed and coastal areas [19, 20].

The output files of SWAN model simulations were matched with the corresponding values obtained from satellite measurements. The matching process was based on the temporal and spatial coordinates of each satellite trajectory. For every altimetry point in the satellite data, the corresponding pixel values of significant wave height (H_s) from the SWAN model was determined and paired accordingly.

3.1 Empirical Wind Speed Estimation from Significant Wave Height

Since the SWAN model does not directly output wind speed, an empirical formula was applied to estimate this parameter, utilizing the significant wave height (H_s) values generated by the simulations. Firstly, the relationship between H_s and U_{10} is expressed through the following equation:

$$H_s\,(U_{10}) \simeq 0.0246 \cdot U_{10}^2 \quad (3)$$

where H_s represents the significant wave height (m), U_{10} is the 10-meter wind speed (m/s) and 0.0246 is an empirical value.

Equation 3 has been determined from classic studies on wind-wave interactions by Pierson-Moskowitz [21] and has been widely adopted and employed by other researchers and engineers in the field [22–24]. Based on the previous Eq. 3, the wind speed can be determined and expressed as:

$$U_{10} = \sqrt{\frac{H_s}{0.0246}} \quad (4)$$

Equation 4 was applied to the H_s values obtained from SWAN model simulations and compared with the 10-meter wind speed satellite values thus providing a fast and practical estimation of wind speed, effective for regions with developed waves, such as the Black Sea.

4 Results

The graphical representations of the temporal variations in daily mean wind speed values (see Figs. 3, 4, 5, 6 and 7) for each satellite reveal wind speeds reaching up to 14 m/s, with occasional peaks exceeding this value, particularly during the winter months. Overall, there are no significant differences in wind speed trends between the satellites.

When examining the daily mean wind speed values across all four satellites combined (Fig. 7), wind speeds frequently surpass 6 m/s, with some instances exceeding 12 m/s.

In contrast, the wind speed estimates from the SWAN model rarely exceed 6 m/s. This suggests that the empirical formula used (Eq. 4) may underestimate higher wind events in the Black Sea region, particularly during extreme conditions. It's important to note that the wind speed estimations rely on the quality of SWAN model simulation outputs on significant wave height (H_s) as indicated in previous studies [25], they may generally be underestimated. Despite the differences between these two datasets, the results on wind speed (U_{10}) consistently reflect similar

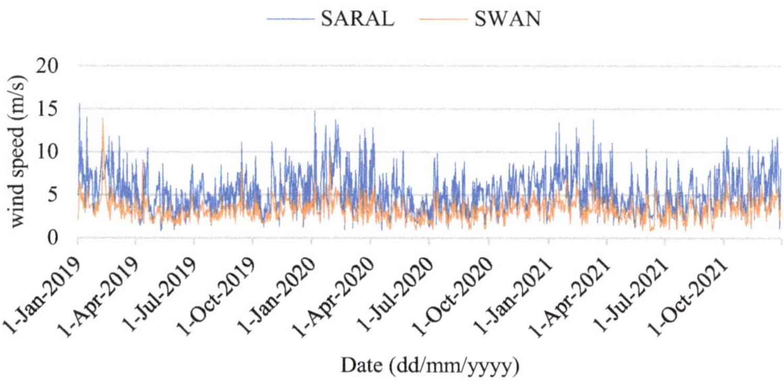

Fig. 3 Temporal variation of daily mean values of wind speed U_{10} (m/s) based on SARAL satellite data and estimations from SWAN model simulations, 01.01.2019–31.12.2021

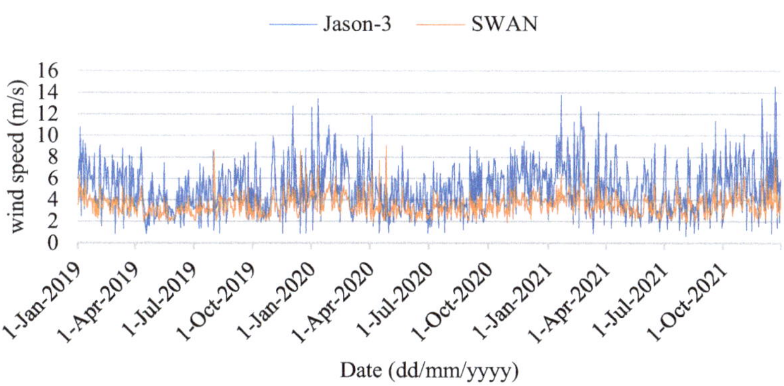

Fig. 4 Temporal variation of daily mean values of wind speed U_{10} (m/s) based on Jason-3 satellite data and estimations from SWAN model simulations, 01.01.2019–31.12.2021

seasonal trends and notable wind events, indicating a robust correlation between the datasets.

The wind roses displaying the distribution of winds by direction and speed based on satellite data (Figs. 8, 9, 10, 11 and 12) indicate that the N and NE directions are dominant in both frequency and velocity. Moderate wind speeds of up to 6.8 m/s occur with a frequency between 10% and 14%, while higher speeds reaching up to 10.8 m/s show a frequency of 6% to 12%, as observed across multiple satellites. In contrast, the calmest wind directions are S, SE, and SW, with frequencies not exceeding 6%, and are associated with weaker wind velocities. The wind directions were derived from the satellite's U and V component values.

The distribution of mean wind directions (Figs. 13, 14, 15, 16 and 17) reveals a clear predominance of winds originating from the N and NE, with frequencies ranging between 15% and 19% across all satellite datasets. In contrast, winds from

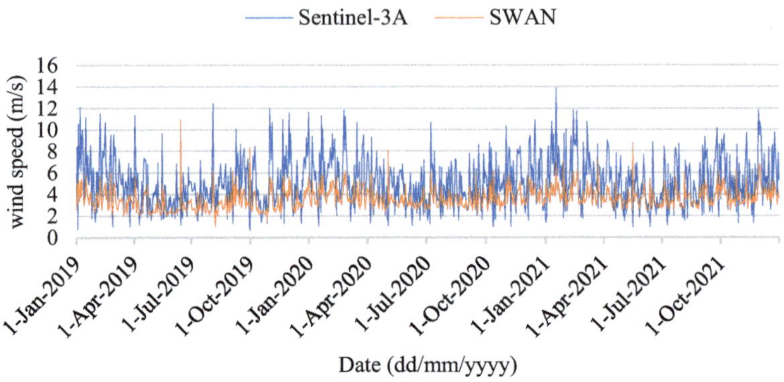

Fig. 5 Temporal variation of daily mean values of wind speed U_{10} (m/s) based on Sentinel-3A satellite data and estimations from SWAN model simulations, 01.01.2019–31.12.2021

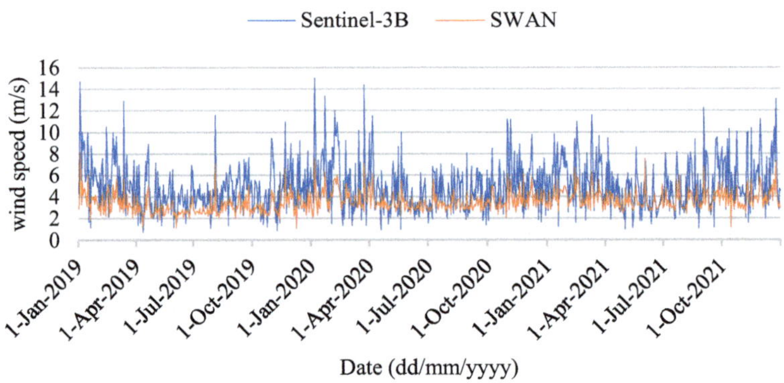

Fig. 6 Temporal variation of daily mean values of wind speed U_{10} (m/s) based on Jason-3 satellite data and estimations from SWAN model simulations, 01.01.2019–31.12.2021

the S, SE, and SW directions exhibit the lowest frequencies, not exceeding 10%. This indicates that northern and northeastern winds are dominant in the study area, both in terms of frequency and, as observed earlier, wind speed, while southern winds are less common.

Assessing wind speed patterns in the Black Sea using satellite altimetry measurements, paired with estimations derived from SWAN model simulation outputs, yields promising results. The results of statistical analysis are summarized in Table 2. The wind speed estimation formula (Eq. 4) is based on significant wave height (H_s) values. It is evident that enhancing the model through the use of an unstructured mesh, fine-tuning physical processes and adjustable parameters, and increasing the resolution of bathymetric data, particularly in coastal regions, could improve the accuracy of U_{10} estimates.

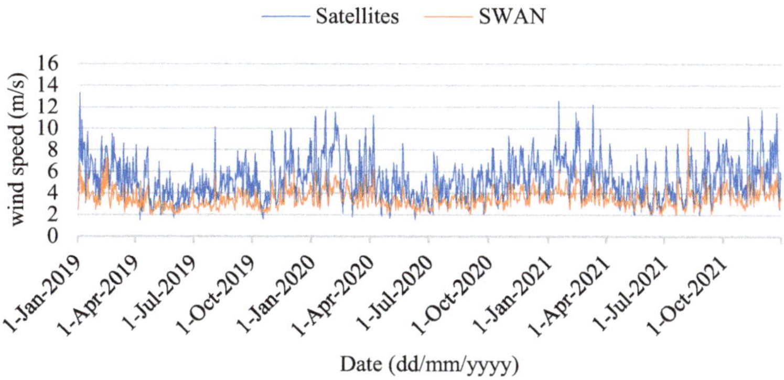

Fig. 7 Temporal variation of daily mean values of wind speed U_{10} (m/s) based on all satellite data and estimations from SWAN model simulations, 01.01.2019–31.12.2021

Fig. 8 Wind direction and wind speed distribution over the Black Sea based on SARAL data, 01.01.2019–31.12.2021

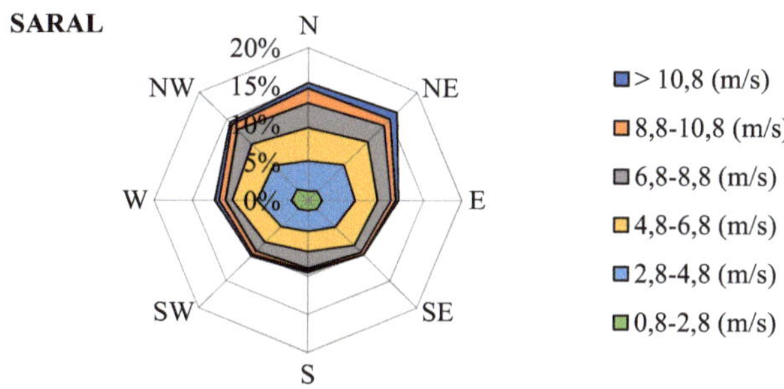

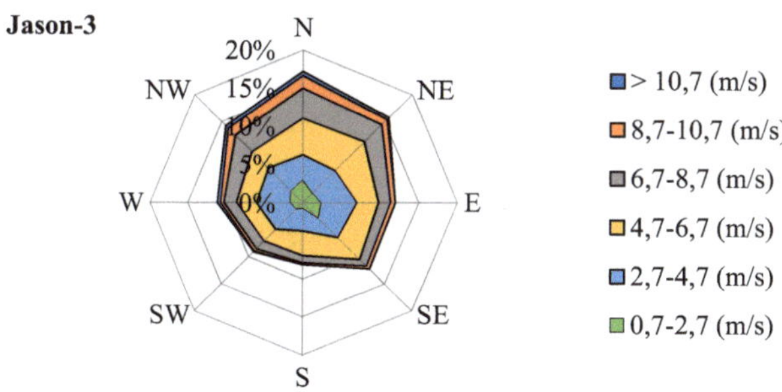

Fig. 9 Wind direction and wind speed distribution over the Black Sea based on Jason-3 data, 01.01.2019–31.12.2021

The RMSE values represent the average deviation between the multiannual mean wind speed values measured by satellite altimetry and those estimated using Eq. 4, which is applied to the significant wave height (H_s) values derived from SWAN model simulations.

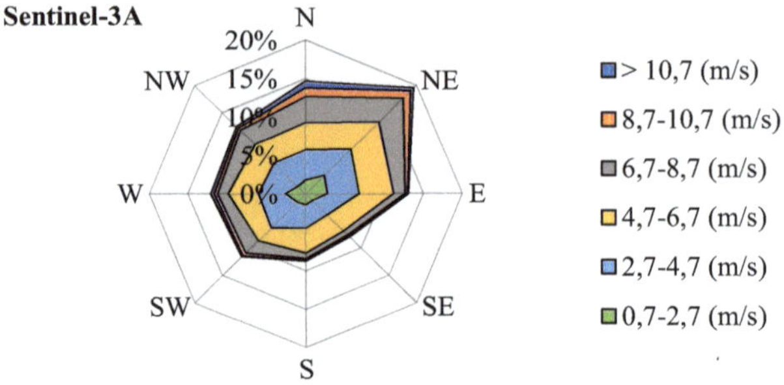

Fig. 10 Wind direction and wind speed distribution over the Black Sea based on Sentinel-3A data, 01.01.2019–31.12.2021

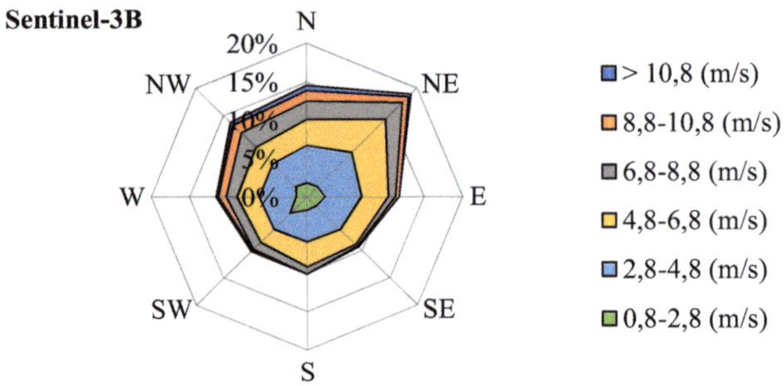

Fig. 11 Wind direction and wind speed distribution over the Black Sea based on Sentinel-3B data, 01.01.2019–31.12.2021

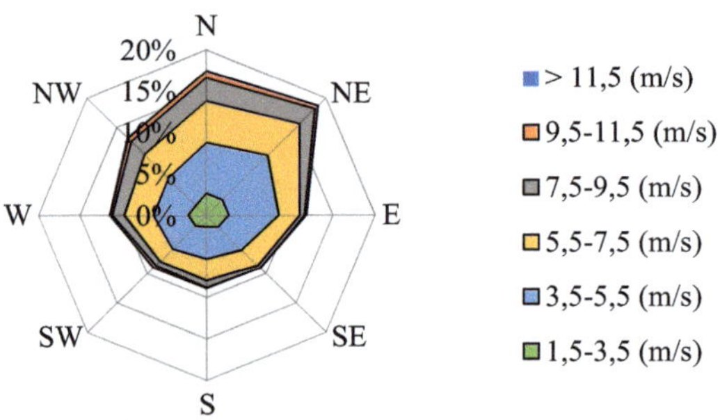

Fig. 12 Wind direction and wind speed distribution over the Black Sea based on data from all four satellites, 01.01.2019–31.12.2021

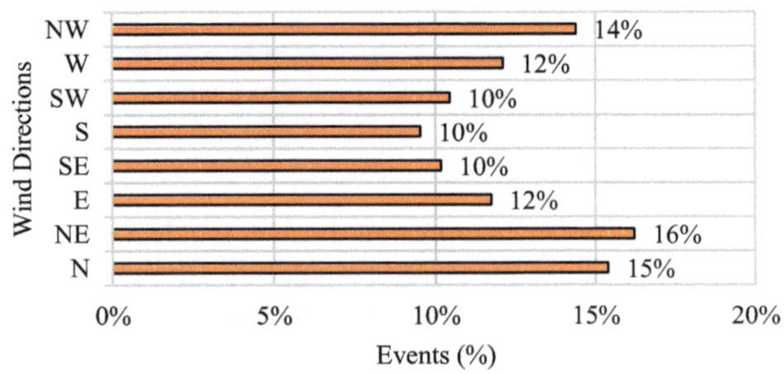

Fig. 13 Distribution of mean wind directions over the Black Sea calculated from SARAL satellite data, 01.01.2019–31.12.2021

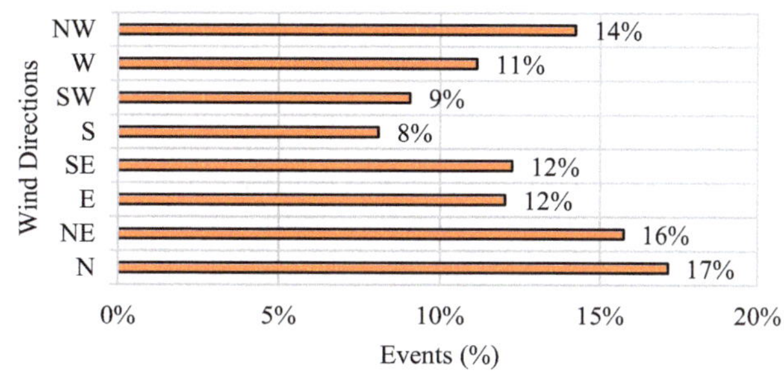

Fig. 14 Distribution of mean wind directions over the Black Sea calculated from Jason-3 satellite data, 01.01.2019–31.12.2021

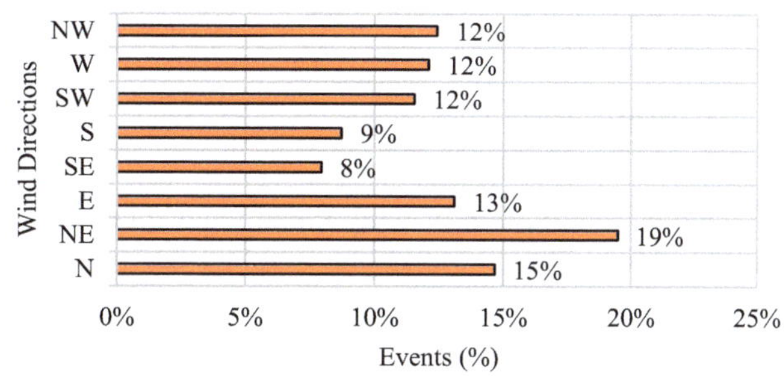

Fig. 15 Distribution of mean wind directions over the Black Sea calculated from Sentinel-3A satellite data, 01.01.2019–31.12.2021

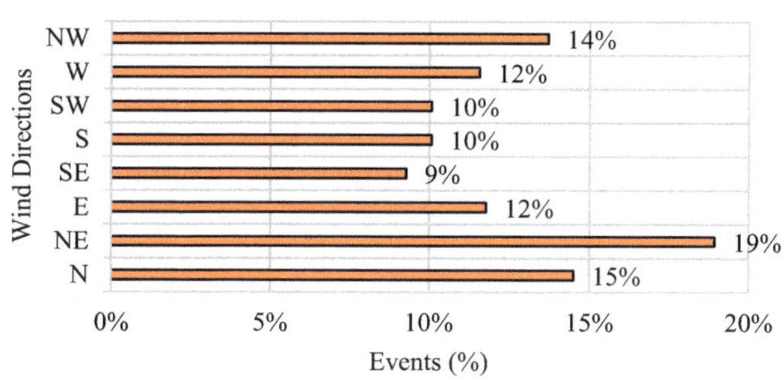

Fig. 16 Distribution of mean wind directions over the Black Sea calculated from Sentinel-3B satellite data, 01.01.2019–31.12.2021

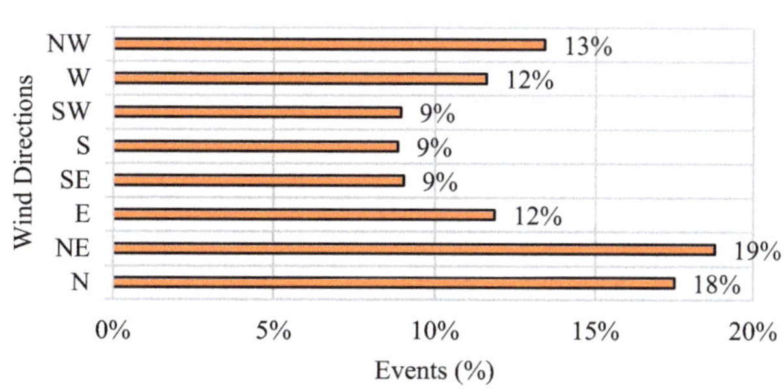

Fig. 17 Distribution of mean wind directions over the Black Sea calculated from all four satellites, 01.01.2019–31.12.2021

Table 2 Statistics on the wind speed (U_{10}) between measured values by satellites and estimated values resulted from SWAN model simulations, on the Black Sea, 01.01.2019–31.12.2021

Satellite	Multiannual mean wind speed values from satellite data [m/s] 2019–2021	Multiannual mean wind speed values from SWAN model simulations [m/s] 2019–2021	Correlation coefficient (R)	Root Mean Square Error (RMSE)
SARAL	5.47	3.37	0.79	2.16
Jason-3	5.16	3.55	0.76	2.28
Sentinel-3A	5.07	3.60	0.80	2.11
Sentinel-3B	5.01	3.52	0.78	2.35
All satellites	5.35	3.51	0.86	2.23

5 Conclusions

The temporal variation of daily mean wind speed (U_{10}) shows a consistent pattern across all four altimetry missions. Satellite-derived wind speed values display higher short-term variability, with peaks exceeding 10 m/s, compared to the estimations derived from applying Eq. 4 to significant wave height (H_s) values produced by the SWAN model.

The wind roses, created from satellite data, confirm that the N and NE directions dominate both in frequency and intensity across all datasets. Moderate winds up to 6.8 m/s are most frequent, while stronger winds above 8.7 m/s are mainly linked with the N and NE directions. By contrast, the S, SE, and SW directions are less frequent and calmer.

The directional frequency distribution confirms the dominance of N and NE winds, with consistent frequencies above 15% for these directions across all missions while S and SE winds remain less common and less significant in shaping the basin's wind regime.

Although potential underestimations may arise from lower-quality input wind fields, the SWAN model simulations continue to provide consistently reliable results for the Black Sea basin. The strong correlation coefficient of 0.86 emphasizes the robustness and accuracy of both satellite altimetry data and SWAN model outputs in characterizing wind-wave dynamics in the region.

References

1. Cavaleri, L., Alves, J.H.G.M., Ardhuin, F., Babanin, A., Banner, M., Belibassakis, K.A., Young, I.R.: Wave modelling—the state of the art. Prog. Oceanogr. **75**(4), 603–674 (2007)
2. Van Vledder, G.P., Akpınar, A.: Wave model predictions in the Black Sea: sensitivity to wind fields. Appl. Ocean Res. **53**, 161–178 (2015)
3. Booij, N., Ris, R.C., Holthuijsen, L.H.: A third-generation wave model for coastal regions: 1. Model description and validation. J. Geophys. Res. Oceans. **104**(C4), 7649–7666 (1999)

4. WAMDI Group: The WAM model—a third generation ocean wave prediction model. J. Phys. Oceanogr. **18**(12), 1775–1810 (1988)
5. Tolman, H.L.: User Manual and System Documentation of WaveWatch III Version 3.14. NOAA/NWS/NCEP/MMAB Technical Note. **276**, 1–220 (2009)
6. Rusu, E.: Strategies in using numerical wave models in ocean/coastal applications. J. Mar. Sci. Technol. **19**(1), 58–75 (2011)
7. Galabov, V., Kortcheva, A., Bogatchev, B., Tsenova, B.: Investigation of the hydro-meteorological hazards along the Bulgarian coast of the Black Sea by reconstructions of historical storms. J. Environ. Prot. Ecol. **16**(3), 1005–1015 (2015)
8. Lee, L.F., Anny, C.: Satellite Altimetry and Earth Sciences—A Handbook of Techniques and Applications, vol. 69. Academic Press, San Diego (2001)
9. Australian Ocean Data Network portal, https://portal.aodn.org.au/search. Accessed 12 July 2024
10. Luca, E., Bandoc, G., Degeratu, M.: Comparative analysis of significant wave height between satellite altimetry data and SWAN model simulations in the Black Sea basin. IOP Conf. Ser. Earth Environ. Sci. **1185**(1), 012022 (2023)
11. Luca, E., Golumbeanu, M.: Analysis of significant wave height for the black sea basin using aviso satellite altimeter data. J. Environ. Prot. Ecol. **21**(2), 405–412 (2020)
12. Sverdrup, H.U., Munk, W.H.: Wind, sea and swell. Theory of relations for forecasting. U.S. Navy Hydrographic Office. **601**(1), 1–44 (1947)
13. SWAN Team 2021 SWAN User Manual/SWAN Cycle III version 41.31AB Delft University of Technology, Netherlands, 1–138
14. Hasselmann, K., Barnett, T.P., Bouws, E., Carlson, H., Cartwright, D.E., Enke, K., Ewing, J.A., Gienapp, H., Hasselmann, D.E., Kruseman, P., Meerburg, A., Müller, P., Olbers, D.J., Richter, K., Sell, W., Walden, H.: Measurements of wind-wave growth and swell decay during the joint North Sea wave project (JONSWAP). Ergänzungsheft zur Deutschen Hydrographischen Zeitschrift. **A8**(12), 95 (1973)
15. Booij, N., Ris, R.C., Holthuijsen, L.H.: a third-generation wave model for coastal regions. Part I: model description and validation. J. Geophys. Res. Oceans. **104**(C4), 7649–7666 (1999)
16. Holthuijsen, L.H.: Waves in Oceanic and Coastal Waters. Cambridge University Press, Cambridge (2007)
17. Tolman, H.L.: A third-generation model for wind waves on slowly varying, unsteady, and inhomogeneous depths and currents. J. Phys. Oceanogr. **28**(4), 479–501 (1998)
18. Chawla, A., Tolman, H.L., Spindler, D.M.: Validation of a thirty year wave hindcast using the climate forecast system reanalysis winds. Ocean Model. **70**, 189–206 (2013)
19. Bingölbali, B., Akpınar, A., Jafali, H., Van Vledder, G.P.: Downscaling of wave climate in the western Black Sea. Ocean Eng. **172**, 31–45 (2019)
20. Amarouche, K., Akpınar, A., Çakmak, R.E., Houma, F., Bachari, N.E.I.: Assessment of storm events along the Algiers coast and their potential impacts. Ocean Eng. **210**, 1–15 (2020)
21. Pierson Jr., J.W., Moskowitz, L.: A proposed spectral form for fully developed wind seas based on the similarity theory of of S. A. Kitaigorodskii. J. Geophys. Res. **69**(24), 5181–5190 (1964)
22. Gao, Y., Schmitt, F., G., Hu, J., Huang, Y.: Probability-based wind-wave relation. Front. Mar. Sci. **9**, 1–16 (2023)
23. Goda, Y.: Directional wave spectrum and its engineering applications. Adv. Coast. Ocean Eng. **3**, 67–102 (1997)
24. Liu, Y., Li, S., Yi, Q., Chen, D.: Wind profiles and wave spectra for potential wind farms in South China Sea. part II: wave spectrum model. Energies. **10**(1), 127 (2017)
25. Myslenkov, S.A., Chernyshova, A.: Comparing wave heights simulated in the Black Sea by the SWAN model with satellite data and direct wave measurements. Russ. J. Earth Sci. **16**(5), 1–12 (2016)

The Evolution of the Atmospheric Boundary Layer During Convective Storms in the Area of the Municipality of Bucharest Based on Sodar Data

Iris Ţiţineanu, Georgeta Bandoc, Mircea Degeratu, and Eduard Luca

Abstract This paper presents the evolution of the wind profile in the atmospheric boundary layer during convective storms in the area of the Municipality of Bucharest. Sodar soundings were used to analyse the variations of the mean wind speed and wind speed shear velocity in the atmospheric boundary layer. The data was collected during, as well as 12 h before and 12 h after the nowcasting warnings for severe thunderstorms issued by the National Meteorological Administration between 2018 and 2021. Thus, the structure and evolution of the wind speed and wind shear velocity profiles in the atmospheric boundary layer was observed prior to, during and following severe convective episodes. The results are important in understanding the processes that take place in the atmospheric boundary layer during convective storms.

Keywords Boundary layer · Wind · Convective storms · Turbulence · Sodar

I. Ţiţineanu (✉)
University of Bucharest, Faculty of Geography, "Simion Mehedinţi" Doctoral School, Center for Coastal Research and Environmental Protection, Bucharest, Romania

Meteo Romania (National Meteorological Administration), Bucharest, Romania

G. Bandoc
University of Bucharest, Faculty of Geography, "Simion Mehedinţi" Doctoral School, Center for Coastal Research and Environmental Protection, Bucharest, Romania

Romanian Association for Wind Engineering, Bucharest, Romania

Academy of Romanian Scientists, Bucharest, Romania

M. Degeratu
Romanian Association for Wind Engineering, Bucharest, Romania

Academy of Romanian Scientists, Bucharest, Romania

Technical University of Civil Engineering Bucharest, Bucharest, Romania

E. Luca
University of Bucharest, Faculty of Geography, "Simion Mehedinţi" Doctoral School, Center for Coastal Research and Environmental Protection, Bucharest, Romania

I. Calotescu, C. I. Coşoiu (eds.), *Proceedings of the Third National Conference on Wind Engineering*, Springer Proceedings in Earth and Environmental Sciences,
https://doi.org/10.1007/978-3-032-09671-5_9

1 Introduction

The atmospheric boundary layer (ABL) is the layer where interactions take place between the earth's surface, which captures most of the incoming solar energy and redistributes it in different forms, and the large scale atmospheric flow, which is driven by this energy [1]. The most important transport process in the ABL is turbulence. In the ABL with dominant convective forcing, which is the case of the so-called atmospheric convective boundary layer (CBL), turbulence is primarily forced either by the heating at the underlying surface or radiative cooling in clouds at the top of the layer, or by both mechanisms [2]. It is known that thunderstorm formation requires three basic ingredients to form: static instability resulting from boundary layer solar heating and cold air at mid levels, sufficient moisture content at low levels, and a lifting mechanism to release the convective available potential energy (CAPE) [3, 4]. Severe convective storms occur in thermodynamically unstable environments where wind shear is high, and there is sufficient lifting capable of triggering convection and releasing the thermodynamic instability [5–7]. In southeastern Romania, severe thunderstorms develop frequently during the convective season (from May to September), and they sometimes lead to significant damages. Good wind speed measurements are necessary in order to identify and examine the properties of the atmospheric boundary layer, such as convective structures or turbulence. Measurements by remote sensing instruments such as Doppler sodar offer advantages over point measurements as they yield wind information from different heights simultaneously [8–12].

The goal of this study is to describe the evolution of the ABL prior to, during and following convective storms, as an effort to understand the processes that take place during these severe weather events.

2 Data and Methods

The sodar (Sonic Detection and Ranging) measures the wind profile by emitting acoustic pulses into the atmosphere and analyzing the backscattered signals. It operates by transmitting acoustic energy into the atmosphere and measuring the strength and frequency of backscattered energy.

This paper used wind speed and wind shear velocity measurements with the METEK PCS.2000–64 monostatic Doppler sodar conducted over a 4 year period, from 2018 to 2021. The sodar is located at the Henri Coandă International Airport in the northern part of Bucharest and it was operated with a vertical resolution of 30 m, a first range gate (minimum height) of 40 m and a potential maximum range of 610 m. The zenith angle of the sound beams was set to 17° and the integration time to 10 minutes [13]. Wind speed refers to how fast the air is moving past a specific point. The sodar measures the zonal (east-west), meridional (north-south) and vertical components of the wind. Wind shear velocity refers to the difference in wind speed between two height ranges, in this case scaled up to 30 meters.

The archive of severe weather nowcasting alerts from the National Meteorological Administration was consulted in order to identify the dates and times of severe thunderstorm warnings issued for the Municipality of Bucharest from 2018 to 2021. A code yellow warning for wind speed is issued for speeds exceeding 50 m/s and up to 70 km/h, while a code orange warning is usually issued for wind speeds between 70 m/s and 90 m/s. However, for convective storms, the warnings can be issued for large water quantities or hail, even if wind speeds do not reach the specified threshold.

Wind speed and wind shear velocity data during the convective storms, 12 h prior to, but also 12 h after the event was then extracted from the sodar based on the identified events and statistically analysed in order to obtain 1 h averages for wind speed and wind shear velocity. We chose 1 h averages because of the average 1 h duration of the weather warnings identified. The results were then used to construct graphs in Microsoft Excel [14]. The goal was to determine the vertical distribution of the mean U wind speed, which is the zonal component of wind, but also the wind shear velocity in the ABL on days with storm warnings, during the convective event, but also 12 hours prior to and following the thunderstorm, in order to observe the evolution of the boundary layer wind profile.

Because Bucharest is a large city, certain warnings issued for thunderstorms in the area did not translate to significant changes in wind speed and wind shear velocity changes in the area where the sodar is placed. Thus, wind speeds measured by the sodar during each thunderstorm warning issued for the municipality of Bucharest between 2018 and 2021 were consulted and out of all the cases we selected the ones where the increase in wind speed during the warnings was relevant. Out of all the available data, 12 cases were selected for the code yellow warnings and 16 cases for the code orange warnings, which were considered to be significant for this study.

To illustrate the evolution of the mean wind speed and wind shear velocity in the ABL before, during and after convective storms as observed in the northern part of Bucharest, two example cases of the vertical wind speed distribution and vertical wind shear velocity distribution are presented for each weather warning colour.

3 Results

3.1 Vertical Distribution of the Mean Wind Speed inside the ABL on Days with Code Yellow Warnings

Measured data on days with code yellow warnings for convective storms were analysed. The analysis consisted of determining the vertical distribution of the local temporal mean U wind speed in the ABL on days with code yellow weather alerts.

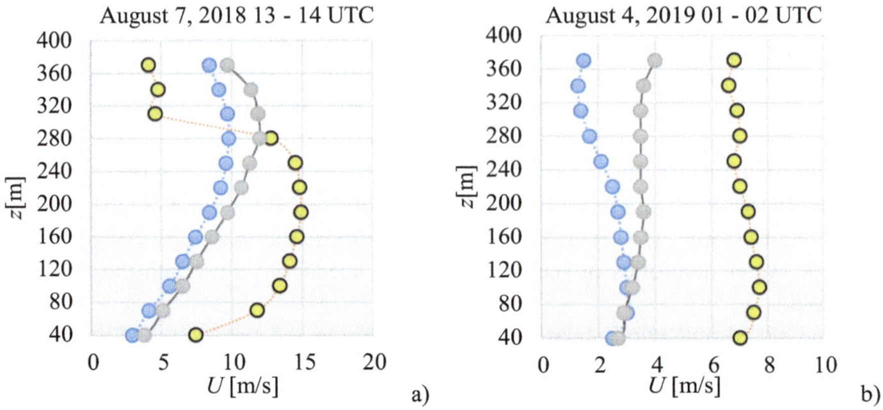

Fig. 1 Variation of wind speed U with height *z* in the ABL of the studied area during code yellow: (**a**) year 2018 and (**b**) year 2019. Wind speeds during the warning are shown in yellow; those 12 hours before are shown in blue, and those 12 hours after in gray

The variations of the mean wind speed U as a function of height z with respect to the ground surface, measured with the sodar equipment, were represented graphically in Fig. 1a, b.

The analysis of the vertical distribution of the mean wind speed in the ABL showed that speeds are generally up to three higher during the code yellow weather warning compared to the mean wind speeds 12 hours before and after the event. This is due to the strong winds that are associated with the downdrafts and outflow boundaries of the storm. As the storm dissipates, the wind speeds diminish, but the atmospheric conditions that led to the storm can persist. As a result, the wind speeds measured 12 hours after the warning generally decrease but remain slightly higher than the speeds determined 12 h before the thunderstorm. This can be observed through the entire depth of the ABL.

3.2 Vertical Distribution of the Wind Shear Velocity inside the ABL on Days with Code Yellow Warnings

Measured data on days with code yellow warnings for convective storms using the sodar were analysed. The analysis consisted of determining the vertical distribution of the wind shear velocity U_1 in the ABL on days with code yellow weather alerts.The variations of the wind shear velocity U_1 as a function of height z with respect to the ground surface, measured with the sodar equipment, were represented graphically in Fig. 2a, b.

The analysis of the vertical distribution of the wind shear speed in the ABL showed great variations between wind shear speeds during, before and after the

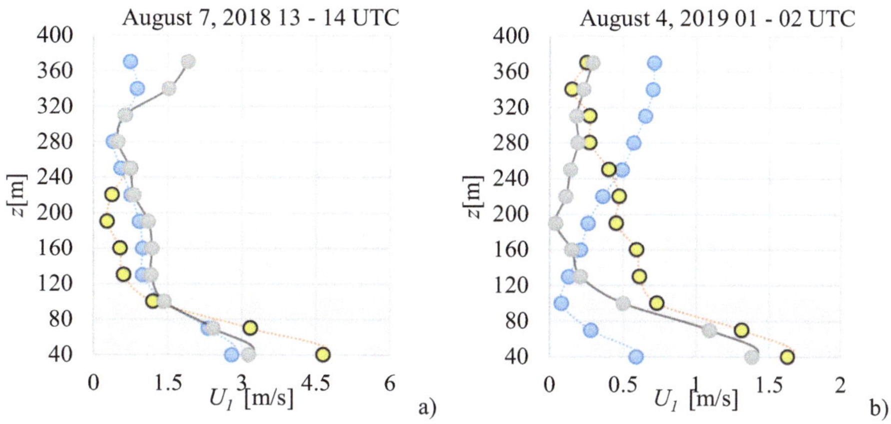

Fig. 2 Variation of shear velocity U_1 with height z in the ABL of the studied area during code yellow: (**a**) year 2018 and (**b**) year 2019. Wind shear velocities during the warning are shown in yellow; those 12 hours before are shown in blue, and those 12 h after in gray

convective storm event, up to around 120 m above ground level. This indicates that the air in this layer participates in an intense turbulent motion. Above the 120 m height and up to the top of the ABL the difference between the wind shear velocities greatly reduces, and reaches similar values to those before the convective event.

3.3 Vertical Distribution of the Mean Wind Speed inside the ABL on Days with Code Orange Warnings

Measured data on days with code orange warnings for convective storms were analysed. The analysis consisted of determining the vertical distribution of the local temporal mean U wind speed in the ABL on days with code orange weather alerts for thunderstorms. The variations of the mean wind speed U as a function of height z with respect to the ground surface, measured with the sodar equipment, were represented graphically in Fig. 3a, b.

The analysis of the vertical distribution of the mean wind speed in the ABL showed that speeds are generally up to five times higher during the code orange weather warning compared to the mean wind speeds 12 h before and after the event. This is a result of convective storms producing strong surface winds, including damaging gusts during their active phase, often associated with the outflow boundaries of the storm which is the area where cool air sinks because of its higher density and spreads out.

The wind speeds measured 12 h after the warning generally decrease but remain slightly higher than the wind speeds determined 12 h before the thunderstorm, and the difference between the two is slightly higher than in the code yellow cases. This

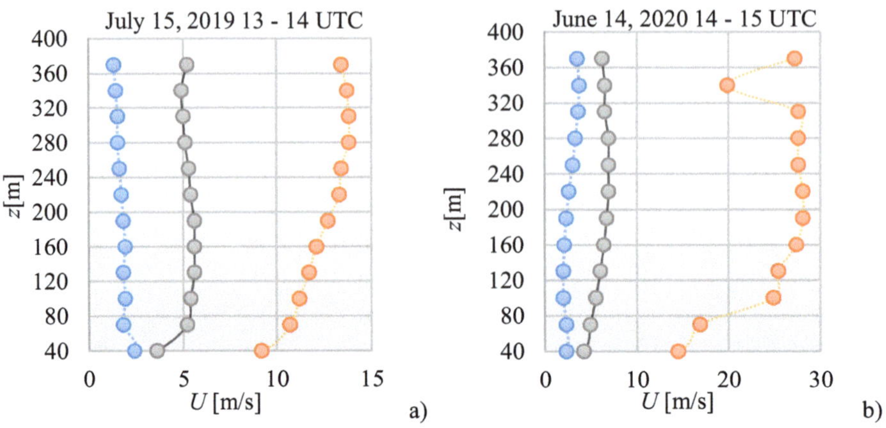

Fig. 3 Variation of wind speed U with height z in the ABL of the studied area during code orange: (**a**) year 2019 and (**b**) year 2020. Wind speeds during the warning are shown in orange; those 12 hours before are shown in blue, and those 12 h after in gray

can be observed through the entire depth of the ABL. This might be a result of the storm weakening or moving away. The updrafts and downdrafts of the storm thus become less vigorous, leading to a drop in wind speed. However, the atmospheric conditions that supported the storm may not immediately dissipate due to the mixing and momentum distribution caused by convection, which may cause the wind speeds after the storm to remain slightly higher than the wind speeds before.

3.4 Vertical Distribution of the Wind Shear Velocity inside the ABL on Days with Code Orange Warnings

Measured data on days with code orange warnings for convective storms using the sodar were analysed. The analysis consisted of determining the vertical distribution of the wind shear velocity U1 in the ABL on days with code orange weather alerts for convective storms. The variations of the wind shear velocity U1 as a function of height z with respect to the ground surface, measured with the sodar equipment, were represented graphically in Fig. 4a, b.

The analysis of the vertical distribution of the wind shear speed in the ABL showed great variations between wind shear speeds during, before and after the convective storm event, up to around 120 m above ground level. This indicates that the air in this layer participates in an intense turbulent motion. Above the 120 m height and up to the top of the ABL the difference between the wind shear velocities greatly reduces, and reaches similar values to those before the convective event.

This is most likely a result of the downdrafts from the storm spreading out near the ground and creating significant changes in wind speed and direction. Another

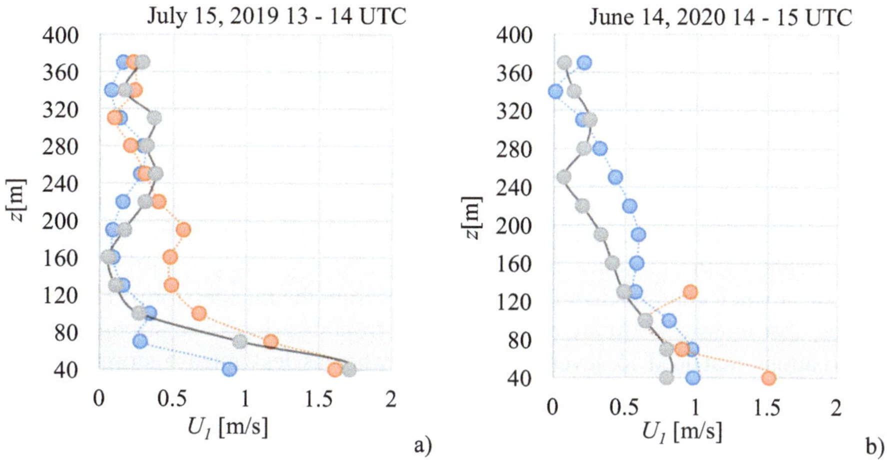

Fig. 4 Variation of shear velocity U_1 with height z in the ABL of the studied area during code orange: (**a**) year 2019 and (**b**) year 2020. Wind shear velocities during the warning are shown in orange; those 12 hours before are shown in blue, and those 12 h after in gray

cause might be the presence of low level jet streams, that can further increase wind shear when combined with the downdrafts from the thunderstorm.

4 Conclusions

The analysis of the vertical distribution of the mean wind speed in the ABL showed that speeds are higher during the code yellow and code orange weather warnings compared to the mean wind speeds 12 h before and after the events.This is a result of convective storms producing strong surface winds, including damaging gusts during their active phase, often associated with the outflow boundaries of the storm which is the area where cool air sinks because of its higher density and spreads out.

The wind speeds measured 12 h after the warning generally decrease but remain slightly higher than the wind speeds determined 12 h before the thunderstorm. The difference between the two is also slightly higher in the code orange cases than in the code yellow cases. This can be observed through the entire depth of the ABL.

This is indicative of the storm weakening or moving away. The updrafts and downdrafts of the storm thus become less vigorous, leading to a drop in wind speed. However, the atmospheric conditions that supported the storm may not immediately dissipate due to the mixing and momentum distribution caused by convection, which may cause the wind speeds after the storm to remain slightly higher than the wind speeds before.

The analysis of the vertical distribution of the wind shear speed in the ABL showed significant variations between wind shear speeds during, before and after the convective storm events, up to around 120 m above ground level due to the increased wind turbulence in an unstable ABL. This shows that the air in this layer participates in an intense turbulent motion resulting from the overlap of turbulence of both dynamic and thermal nature. Above the 120 m height and up to the top of the ABL the difference between the wind shear velocities greatly reduces, and reaches similar values to those before the convective events.

This is most likely a result of the downdrafts from the storm spreading out near the ground and creating significant changes in wind speed and direction. Another cause might be the presence of low level jet streams, that can further increase wind shear when combined with the downdrafts from the thunderstorm. The results indicate that intense storm are associated with frontal systems and strong pressure perturbations that lead to severe wind gusts, whereas less intense storms do not disturb the boundary layer as much.

Acknowledgements We would like to thank ROMATSA (Romanian Air Traffic Services Administration) for providing us with the sodar data that this study was based on.

References

1. Haggagy, M.E.N.A. A sodar-based investigation of the atmospheric boundary layer. Dissertation, Albert-Ludwigs-Universität Freiburg, (2003)
2. Conzemius, R.J., Fedorovich, E.: Dynamics of sheared convective boundary layer entrainment. Part I: methodological background and large-eddy simulations. J. Atmos. Sci. **63**(4), 1151–1178 (2006)
3. Kolendowicz, L., Taszarek, M., Czernecki, B.: Atmospheric circulation and sounding-derived parameters associated with thunderstorm occurrence in Central Europe. Atmos. Res. **191**, 101–114 (2017)
4. Doswell III, C.A., Schultz, D.M.: On the use of indices and parameters in forecasting severe storms. E-J. Severe Storms Meteorol. **1**(3), 1–22 (2006)
5. Ribeiro, B.Z., Bosart, L.F.: Elevated mixed layers and associated severe thunderstorm environments in south and North America. Mon. Weather Rev. **146**(1), 3–28 (2018)
6. Brooks, H.E., Lee, J.W., Craven, J.P.: The spatial distribution of severe thunderstorm and tornado environments from global reanalysis data. Atmos. Res. **67–68**, 73–94 (2003)
7. Thompson, R.L., Edwards, R., Hart, J.A., Elmore, K.L., Markowski, P.: Close proximity soundings within supercell environments obtained from the rapid update cycle. Wea. Forecasting. **18**, 1243–1261 (2003)
8. Hooper, W.P., Eloranta, E.W.: Lidar measurements of wind in the planetary boundary layer: the method, accuracy and results from joint measurements with radiosonde and kytoon. J. Appl. Meteorol. Climatol. **25**(7), 990–1001 (1986)
9. May, P.T., Wilczak, J.M.: Diurnal and seasonal variations of boundary-layer structure observed with a radar wind profiler and RASS. Mon. Weather Rev. **121**(3), 673–682 (1993)
10. Lang, S., McKeogh, E.: LIDAR and SODAR measurements of wind speed and direction in upland terrain for wind energy purposes. Remote Sens. **3**(9), 1871–1901 (2011)
11. Ţiţineanu, I., Bandoc, G., Degeratu, M.: A study on the characteristics of the atmospheric boundary layer using SODAR measurements. IOP Conf. Ser. Earth Environ. Sci. **1185**, 012023 (2023). https://doi.org/10.1088/1755-1315/1185/1/012023
12. Ţiţineanu, I., Bandoc, G., Degeratu, M.: An analysis of wind speed variations inside and outside of the atmospheric boundary layer during convective storms using SODAR and

radiosonde data. Annals of the Academy of Romanian Scientists, series Engineering Sciences. **16**(nr.1/2024), 109–117 (2024). https://doi.org/10.56082/annalsarscieng.2024.1.109
13. Buzdugan, L., Stefan, S.: A comparative study of sodar, lidar wind measurements and aircraft derived wind observations. Rom. J. Phys. **65**, 810 (2020)
14. XLSTAT: Data Analysis and Statistical Solution for Microsoft Excel. Addinsoft, Paris, France (2017)

Measurements of Longitudinal Velocity Distributions in a Boundary Layer Wind Tunnel

Ioel Giger, Costin Ioan Coşoiu, and Andrei Mugur Georgescu

Abstract This paper presents an experimental study conducted in the TASL1-M boundary layer wind tunnel (BLWT) at the Constantin Iamandi Aerodynamics and Wind Engineering Laboratory (LAIV) of the Technical University of Civil Engineering Bucharest (UTCB), Romania. The experimental campaign was conducted to determine the longitudinal velocity distributions along a vertical line placed in the median longitudinal plane of the experimental section. The measurements were performed using a Constant Temperature Anemometer (CTA). The main objective is to simulate wind speed distribution, turbulence intensity, and power spectral density at scale and to match the parameters found in wind codes. The paper describes the experimental setup, including the wind tunnel specifications, the variable roughness system, turbulence generators, and measurement equipment. Two sets of measurements were conducted for different roughness heights. The results obtained for a zero-roughness height and no turbulence generators were compared to theoretical predictions given by the law of the wall. The other results were compared with the Eurocode standard, with non-dimensional longitudinal velocity distributions and turbulence intensity profiles expressed as a function of non-dimensional height. For each exposure factor, defined here as the representation of upstream roughness and flow characteristics in the wind tunnel, the turbulence structure of the simulated wind was emphasized by comparing the power spectral density with the code specifications. The results show good agreement between measured profiles and code specifications, confirming the system's ability to quickly generate realistic boundary layers for various exposure factors.

Keywords BLWT · CTA · The law of the wall

I. Giger · C. I. Coşoiu (✉) · A. M. Georgescu
Aerodynamics and Wind Engineering Laboratory, Technical University of Civil Engineering of Bucharest, Bucharest, Romania
e-mail: costinioan.cosoiu@utcb.ro

I. Calotescu, C. I. Coşoiu (eds.), *Proceedings of the Third National Conference on Wind Engineering*, Springer Proceedings in Earth and Environmental Sciences,
https://doi.org/10.1007/978-3-032-09671-5_10

1 Introduction

Boundary Layer Wind Tunnels (BLWT) are typically used to replicate experimentally, in a laboratory environment, at scale, the characteristics of synoptic wind flow in the atmospheric boundary layer (ABL) [1]. The ABL's mean longitudinal wind speed distribution typically follows a logarithmic profile, similar to flow over a flat plate, which is often used as an approximation [2]. For a smooth, flat surface, the non-dimensional law of the wall expression is well established [3]. Comparable results are expected in a BLWT without roughness elements on its floor.

The BLWT must reproduce at scale in the experimental zone the velocity profile, turbulence intensity distribution, and power spectral density as specified in codes and standards, such as EN 1991-1-4:2005 [4], when investigating the influence of natural wind on structures. Accurate reproduction of these characteristics ensures that the data measured in the wind tunnel is reliable for structural design and analysis.

By placing roughness elements on the wind tunnel floor and turbulence generators upwind of the boundary layer development zone, the flow parameters are modified in the experimental zone, reflecting changes in upwind surface conditions [5]. Different exposure factors need different configurations of the roughness elements, which may involve variations in their shape, height, density, or distribution [6]. Changing these between two tests that require wind blowing over different terrain categories extends the experiment with dead runtime that is not used for measurements [7]. An automatic roughness system significantly reduces the setup time. Moreover, it allows the study of specific effects given by different orographic surface patterns apart from the standardized exposure factors defined in codes and procedures [8].

Different techniques are used to measure velocity profiles. In many studies, multihole probes, such as TFI Cobra probes [9], are employed due to their robustness in terms of reliability and operational speed [10]. They allow for measuring all velocity components, and their acquisition rate is relatively high, at approximately 2 kHz [11]. This allows for the determination and quantification of the flow's turbulence properties. Although the technique is invasive (i.e., measuring velocity at a point requires placing a solid instrument in that position), its advantages make it an optimal choice.

Another technique for velocity measurement is Laser Doppler Velocimetry (LDV). Its main advantage is that it is noninvasive, meaning the flow remains undisturbed by the measurement instrument [12]. It can also measure all three velocity components, although the data rate decreases for coincident measurements of each velocity component that is added up [13]. Furthermore, maintaining a constant data acquisition rate is challenging because each measurement must be validated [14]. Additionally, in an open-circuit wind tunnel such as TASL1-M, achieving a high seeding density, required to ensure the acquisition rate needed for accurate power spectral density estimation, is difficult [15]. Consequently, this technique is rather unsuitable for turbulence measurements, at least in open-circuit boundary layer wind tunnels.

A third laboratory technique for velocity measurements is Constant Temperature Anemometry (CTA). Despite being an invasive method with probes highly sensitive to mechanical stress and accidental impact, it remains unparalleled for turbulence measurements due to its exceptionally high and constant data acquisition rate, reaching up to 250 kHz [16]. The CTA uses a controlled thermal-electrical feedback mechanism to regulate sensor temperature. An auxiliary temperature probe must be used to correlate voltage with velocity and compensate for temperature variations accurately. Due to its high sensitivity, before each experiment, a calibration procedure is needed to be performed. It can simultaneously measure the three velocity components at a very high data rate. Still, the calibration process may be relatively complex without an automatic calibration system [17].

This paper presents longitudinal velocity measurements performed using a CTA in a BLWT. The results were compared to Eurocode standards. To modify the exposure factors, the wind tunnel is equipped with an automatic roughness system that can rapidly change the roughness characteristics of the boundary layer development zone.

2 Experimental Setup and Procedures

2.1 The TASL-1 M Boundary Layer Wind Tunnel

Measurements were conducted in the downwind experimental section of the TASL1-M BLWT at LAIV-UTCB [18]. The wind tunnel has an open-circuit design with a total length of 27 m (see Fig. 1). It is powered by a variable-speed, 200 kW axial fan mounted downstream of the experimental section, which generates airflow by drawing air through the tunnel. As a result, the wind tunnel operates at sub-atmospheric pressure. The velocity is continuously adjustable from 0.5 m/s to 30 m/s.

Air enters the tunnel through a square honeycomb structure with a characteristic length of 3.04 m, which aligns the flow with the tunnel's median longitudinal plane and eliminates the rotational effects induced by the fan. After passing through the honeycomb, the airflow enters a convergent nozzle that reduces the cross-sectional area by a factor of 3.02. Following this contraction, there is a short section where turbulence-reducing nets can be inserted if needed.

Downwind of the contraction, the main portion of the wind tunnel begins, consisting of three sections: the upwind experimental section, the boundary layer development zone, and the downwind experimental section. The tunnel maintains a constant square cross-section, with a side of 1750 mm throughout these sections. The longitudinal length of the downwind experimental section is 3170 mm. A

Fig. 1 TASL1-M Boundary Layer Wind Tunnel

turntable with a diameter of 1600 mm is installed here to set different attack angles between the incoming wind and the models.

Upstream of the downwind experimental section lies the boundary layer development zone, with a fetch length of 15,050 mm. This section is equipped with a variable roughness system composed of 560 bricks, each with a horizontal area of 100 mm × 54 mm and an adjustable vertical height ranging from 0 to 200 mm in 1 mm increments (Fig. 2). The footprint of the roughness elements corresponds to the original design of the wind tunnel floor and cannot be modified [18].

The roughness elements are grouped into sets of 40, distributed across 14 equal rectangular sections, each 1075 mm wide (aligned with the flow direction) and extending across the wind tunnel's entire cross-section. The height of the bricks in each segment can be adjusted independently of other segments using a stepper motor placed beneath the wind tunnel floor, which moves a horizontal platform on which the bricks are mounted.

The upwind experimental section, measuring 2150 mm long, is located upstream of the variable roughness system. Five 1 m tetrahedral spires and a castellated barrier are installed here to increase turbulence when models are placed in the downwind experimental section.

Two Pitot-static probes are positioned upstream of each experimental section to monitor velocity. The Pitot-static probe at the contraction's exit measures velocity at the center of the cross-section. The probe located upstream of the downwind experimental section, is installed at a reference height of 600 mm above the wind tunnel floor. A temperature probe is also permanently installed to account for variations in density and viscosity resulting from temperature changes.

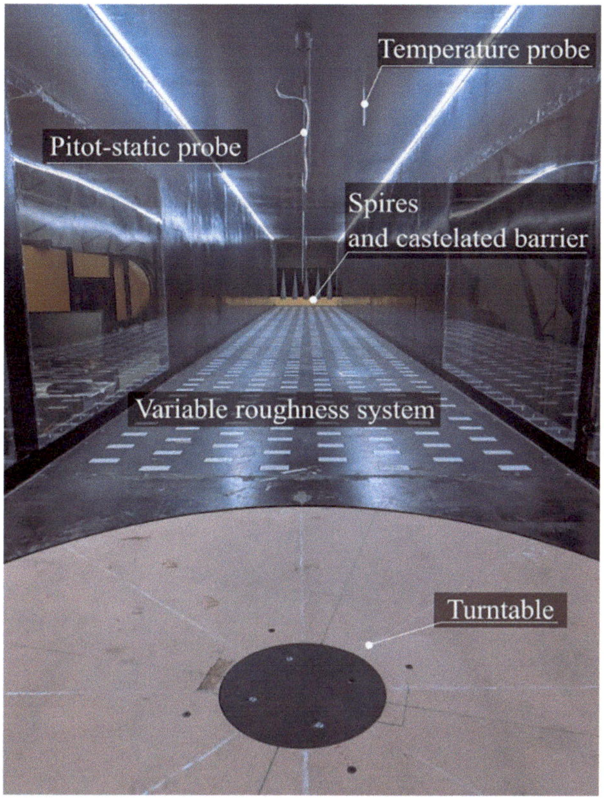

Fig. 2 Interior of the TASL1-M BLWT. View from the downwind experimental section

2.2 Instrumentation and the Measurement Plan

The longitudinal velocity components were measured using a 55P16 1D wire probe connected to a Dantec StreamLine Pro CTA system. A 90P10 temperature probe was also connected to the measurement system to compensate for inherent temperature variations during the experiments. Before each experiment, a calibration procedure was performed using a Dantec automatic calibrator.

The velocity wire probe was mounted on a rigid support fixed to the traversing system inside the downwind experimental section of the TASL1-M BLWT. The traversing system moved the wire probe along the vertical axis, at the center of the turntable, and covered a range from 5 to 600 mm above the wind tunnel floor.

Two sets of measurements were conducted. In the first set (Set 1), the spires and turbulence generators were removed from the wind tunnel, and the roughness height (*RH*) of the roughness elements was set to 0 mm, simulating a flow over a flat plate. The probe was moved between consecutive measurement points with a vertical step Δz of 1 mm from 5 to 200 mm. Over 200 mm and up to 600 mm, the step size

Table 1 Roughness Height (*RH*) in Set 2 Measurements [4]

Terrain category	Description	*RH* [mm]
Cat 0	Sea or coastal area exposed to the open sea	0
Cat I	Lakes or flat and horizontal area with negligible vegetation and without obstacles	20
Cat II	Area with low vegetation, such as grass and isolated obstacles (trees, buildings), with separations of at least 20 obstacle heights	50
Cat III	Area with regular cover of vegetation or buildings, or with isolated obstacles, with separations of a maximum 20 obstacle heights (such as villages, suburban terrain, permanent forest)	72
Cat IV	Area in which at least 15% of the surface is covered with buildings and their average height exceeds 15 m.	96

Table 2 Vertical Coordinates of Measurement Points for Set 2 Measurements

Point #	z [mm]	z/z_{max} [−]
1	7	0.01
2	12	0.02
3	20	0.03
4	33	0.06
5	55	0.09
6	91	0.15
7	150	0.25
8	247	0.41
9	407	0.68
10	600	1.00

was increased to 5 mm. The velocity signal was sampled at a frequency f_s equal to 1 kHz, with an acquisition time (*T*) of 60 s per point. A total of 276 measurement points were considered on Set 1.

The second set of measurements (Set 2) corresponds to five configurations simulating the terrain category effects specified in Eurocode EN 1991-1-4:2005. To replicate these effects at the model scale, the *RH* value was adjusted to different heights, chosen identical across all 14 system segments, as shown in Table 1.

The selected roughness heights were chosen heuristically, based on prior experience with the TASL1-M BLWT for similar experimental configurations.

Velocity was measured at 10 points for each exposure factor. Table 2 presents the vertical coordinates of the measurement points in both dimensional (z) and dimensionless form (z/z_{max}), where z_{max} denotes the maximum vertical coordinate.

The vertical coordinates were distributed according to an exponential law. The first measurement point was 5 mm above the wind tunnel floor, while the last was 600 mm above. The acquisition rate f_s for the second set of measurements was increased to 4 kHz, and the velocity sampling duration T was doubled to 120 s per point. A total of 50 measurements were performed on Set 2, 10 per each exposure factor.

For all measurements performed, both for Set 1 and Set 2, the gradient wind velocity, measured at the vertical coordinate located at z_{max}, was set at approximately 6.5 m/s, corresponding to 20% of the wind tunnel fan's maximum rotational speed.

3 Results

3.1 Set 1 of Measurements

The first set of measurements consists of one case corresponding to a flow over a smooth flat plate upstream of the experimental section. The value of *RH* was set to 0 mm, and the spires and castellated barrier upwind of the boundary layer development zone were removed. As explained in the previous section, the spatial resolution of this particular set of measurements was high, enabling the collection of accurate data for comparison with theoretical findings.

The velocity distribution in a flow over a flat plate follows a logarithmic distribution. When written in nondimensional coordinates, it is described by the following equation, known as the law of the wall [19]:

$$u^+ = \frac{1}{k}\ln\left(z^+\right) + B \tag{1}$$

where $u^+$is the dimensionless velocity given by the ratio between the friction velocity u_* and the longitudinal component of the velocity u:

$$u^+ = u^*/u \tag{2}$$

The friction velocity is given by the formula:

$$u^* = \sqrt{\tau/\rho} \tag{3}$$

where τ is the wall shear stress and ρ is the air density.

In Eq. 1, z^+ is the non-dimensional height, which is defined as the Reynolds number computed using the friction velocity u_* and the distance from the wall to the measuring point:

$$z^+ = \frac{\rho z u^*}{\mu} \tag{4}$$

where μ is the dynamic viscosity of the air.

In the law of the wall, *k* is the von Kármán constant, which is approximately equal to 0.41. *B* is a constant that depends on the surface roughness, and for a smooth, flat plate, it is known to vary between 5 and 5.4 [19].

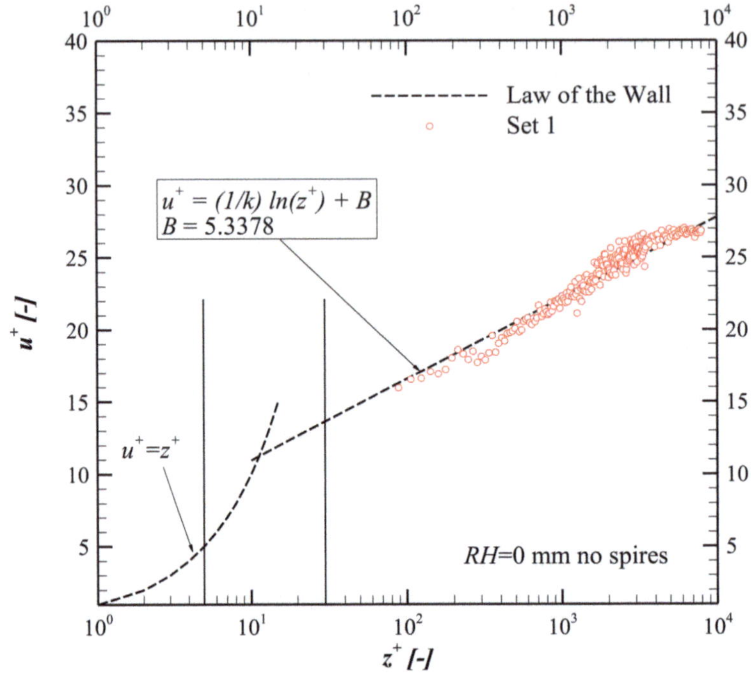

Fig. 3 Comparison of the variation of non-dimensional velocity (u^+) function of non-dimensional height (z^+) corresponding to Set 1 of measurements with theoretical solutions of the law of the wall in the turbulent boundary layer

This variation of the nondimensional velocity with respect to the nondimensional height, as depicted in eq. (1), is known to be true on an interval with values of z^+ higher than 30 and up to approximately 1000. For values of z^+ that varies between 0 and 5, in the so-called laminar sub-layer, the law of the wall is linear so that $u^+ = z^+$. For the region of the buffer layer, situated between z^+ values higher than 5 and lower than 30, neither logarithmic nor linear relations are valid.

Using linear regression applied to a logarithmic transformation of the dimensional log-law equation for velocity variation in the boundary layer, using data from Set 1, resulted in a value of $B = 5.3378$, which falls within the range predicted by theoretical findings.

Figure 3 presents a comparison between the measured data and the theoretical law of the wall written using the found value for the B constant. The variation of the non-dimensional velocity with z^+ closely follows the theoretical findings up to $z^+ \approx 2000$, which corresponds to a dimensional z of approximately 100 mm above the wind tunnel floor. A slight deviation is observed for z^+ varying between 2000 and 5000, where the measured u^+ values are marginally higher than the ones predicted by the theoretical law. For $z^+ > 5000$, the measured data are returning to a logarithmic trend, suggesting that the near-wall model remains a reasonable approximation even at higher wall-normal distances.

3.2 Set 2 of Measurements

As explained in Sect. 2.2 of the present paper, Set 2 of measurements was performed for five values of the roughness height (*RH*), corresponding to the five terrain categories described in Eurocode EN 1991-1-4:2005. The geometrical scale was imposed to be equal to 1:200. Thus, the measurement height of 600 mm of the simulated boundary layer, where velocities were measured in the wind tunnel, corresponds to a height of 120 m at real scale.

Figures 4a–e show the variation of non-dimensional velocity distribution profiles as a function of non-dimensional height for the five considered exposure categories. The measured data, represented by circles, are superimposed on the logarithmic profiles defined in the Eurocode, shown as continuous lines. To better highlight the agreement between measurements and code profiles, the results are plotted on semi-logarithmic axes.

For all exposure categories, the measured velocity profiles slightly overestimate the Eurocode predictions near the wall, particularly for $z/z_{max} < 0.1$. For Cat II, III, and IV, a deviation is also observed in the upper part of the profile, where the measured velocities tend to fall below the code-defined curves. These deviations may be attributed to local flow development in the wind tunnel and to the effect of increased surface roughness, which alters the velocity gradient near the wall and influences the shape of the velocity profile across the full height.

The velocity measurement errors ε_u were decomposed into three components: statistical sampling error ($\varepsilon_{sampling}$), calibration error (ε_{calib}), and drift error (ε_{calib}) [20]. These error components are assumed to be statistically independent and are combined in quadrature:

$$\varepsilon_u = \sqrt{\varepsilon_{sampling}^2 + \varepsilon_{calib}^2 + \varepsilon_{drift}^2} \quad (5)$$

For the Set 2 dataset, the maximum value for the total velocity measurement error ε_u was 1.34×10^{-2} m/s, corresponding to velocity measurements taken at the vertical position $z/z_{max} = 1$. This value represents a relative error of approximately 0.2%, which is well within the lower end of the expected range for such measurements.

Figures 5a–e show the turbulence intensity (*TI*) profiles. The height is also presented in non-dimensional coordinates to match the representation of the velocity distribution. The experimental data aligns well with the code prescriptions. For terrain of Category III and IV, in the lower part of the boundary layer profile (approximately the lowest 10–15% of the simulated ABL), the turbulence intensity values obtained from the experimental data are slightly lower than those prescribed by the code. Although the code prescribes values of 36% for Category IV and 33%

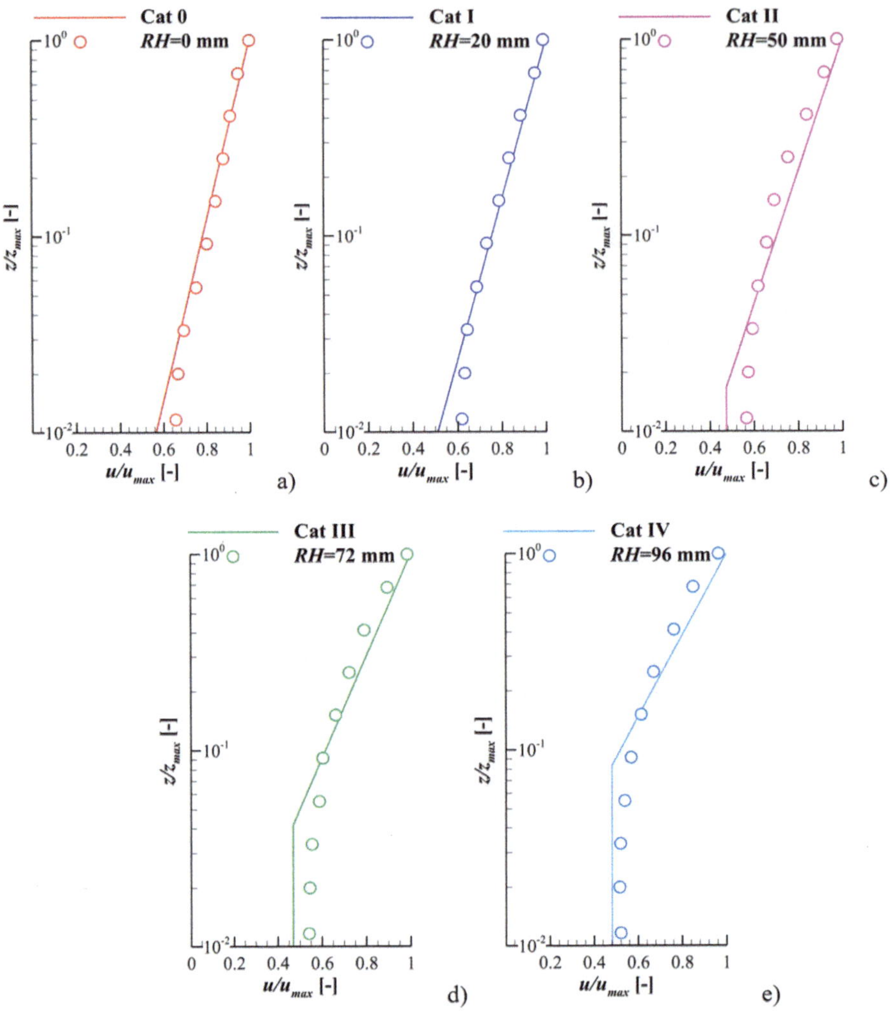

Fig. 4 Nondimensional velocity distribution profiles function of non-dimensional height for the five exposure factors—Set 2 of measurements (max. Total velocity error $\varepsilon_u = 1.34 \times 10^{-2}$m/s i.e. ~ 0.2% at $z/z_{max} = 1$). (**a**) $RH = 0$ mm, (**b**) $RH = 20$ mm, (**c**) $RH = 50$ mm, (**d**) $RH = 72$ mm, (**e**) $RH = 96$ mm

for Category III for turbulence intensity, the maximum values obtained in the wind tunnel are 25% and 22%, respectively. To a lesser extent, the underestimation of the turbulence intensity values in the lower part of the profile is also present for Categories I and II.

The errors in turbulence intensity ε_{TI} were computed as using:

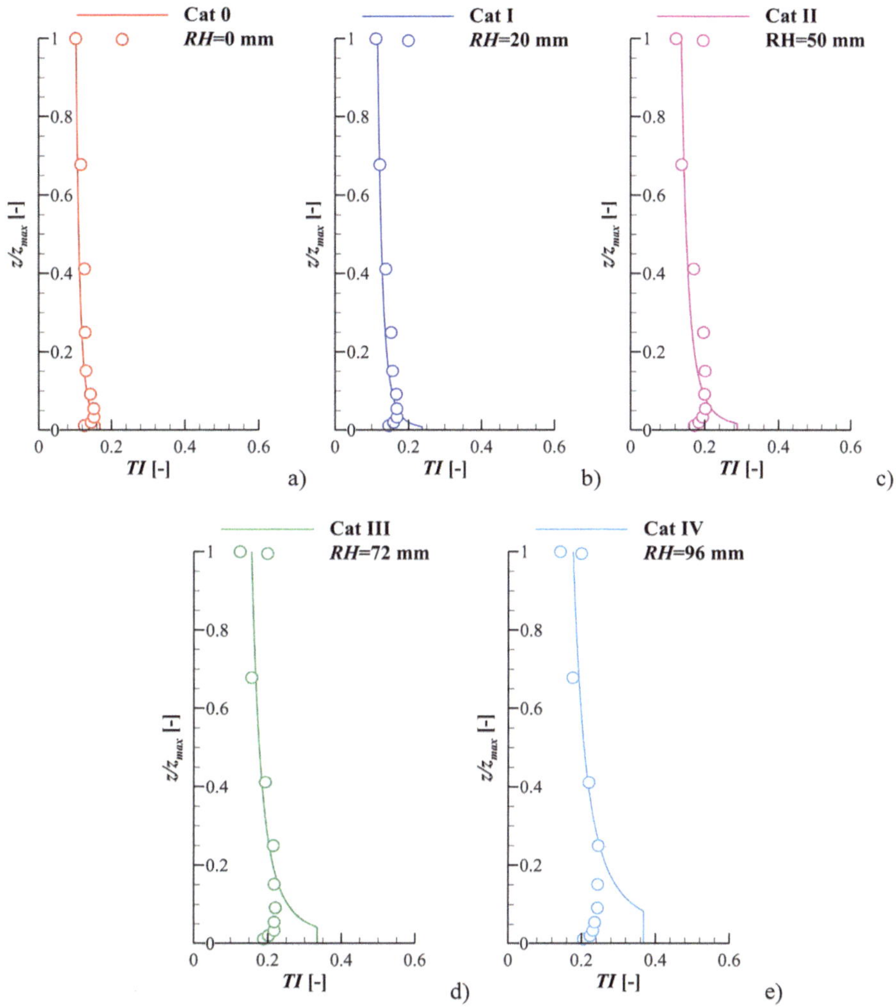

Fig. 5 Turbulence intensity profiles function of non-dimensional height for the five exposure factors—Set 2 of measurements (max. Total turbulence intensity error $\varepsilon_{TI} \approx 1.49 \times 10^{-3}$ i.e. ~ 1.3% at $z/z_{max} = 1$). (**a**) $RH = 0$ mm, (**b**) $RH = 20$ mm, (**c**) $RH = 50$ mm, (**d**) $RH = 72$ mm, (**e**) $RH = 96$ mm

$$\varepsilon_{TI} = TI\sqrt{\left(\frac{\varepsilon_{\sigma u}}{\sigma_u}\right)^2 + \left(\frac{\varepsilon_u}{u}\right)^2} \tag{6}$$

where $\varepsilon_{\sigma u}$, is the standard error of the velocity fluctuation standard deviation, $\sigma_u$is the standard deviation of the instantaneous velocity signal, and the other terms are as previously defined.

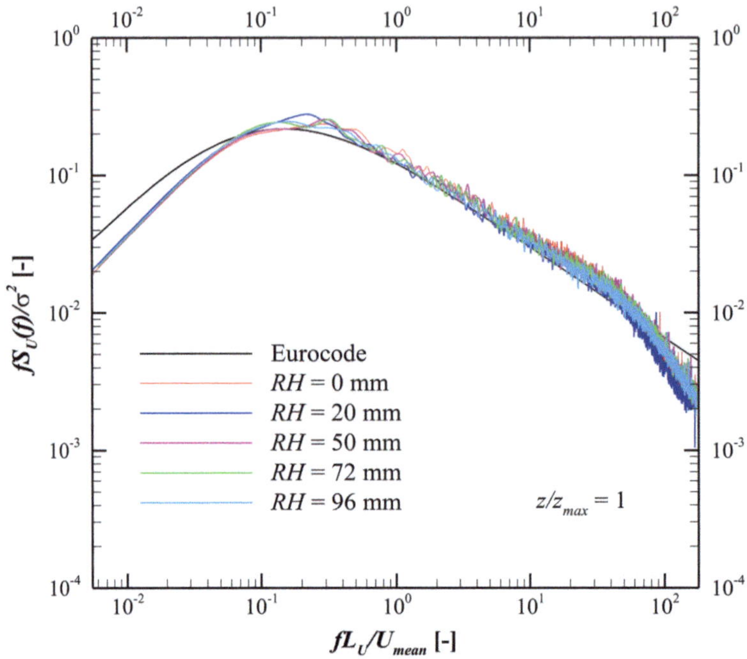

Fig. 6 Power Spectral density function for all five exposure factors at a height corresponding to $z/z_{max} = 1$—Set 2 of measurements

The maximum observed error in turbulence intensity was approximately equal to 1.49×10^{-3}, corresponding to a relative error of about 1.3%, indicating a low measurement uncertainty.

Figure 6 shows the non-dimensional power spectral density for all five exposure categories at a non-dimensional height of $z/z_{max} = 1$, which corresponds to the top-most measurement level. This position, located outside the near-wall region, allows for a more meaningful comparison of the turbulence spectrum across different exposure categories, as it is less affected by surface-induced flow disturbances. The experimental results show excellent agreement with the code prescriptions across all frequencies, irrespective of the set roughness height *RH*, with the exception of the very low-frequency range, where the agreement remains good.

The experimental results show a good agreement between the velocity profiles obtained in the wind tunnel and the prescriptions provided by the Eurocode. The roughness height (*RH*) values used in the simulations were determined heuristically. Thus, it is likely that an even better match with the code could be achieved by adjusting these values further.

The turbulence intensity profiles also align closely with the code-defined values. However, for higher terrain categories, a deviation is observed near the tunnel floor, where the predicted values are smaller than those given by the code. The turbulence intensity increases at the near-wall level rather than being constant, and then decreases as expected according to the code. The underestimation values may be attributed to a low Reynolds number, computed using the local velocity. The tests were performed at a mean flow velocity of approximately 6.5 m/s, measured at a location corresponding to $z/z_{max} = 1$, which may not be sufficient to fully develop a correct turbulent structure of the flow near the wall. Increasing the mean velocity would raise the overall turbulence kinetic energy in the flow, potentially raising the generation of turbulent structures in the near-wall region and resulting in turbulence intensity levels closer to those prescribed by the code. Increasing the mean velocity was not pursued due to practical constraints in the CTA calibration system, which would have required a higher-capacity air supply to extend the calibration curve to higher velocities. Furthermore, at higher flow velocities, internal heating caused by fan operation and flow resistance leads to temperature variations within the tunnel, which, despite compensation, can introduce additional uncertainty in the CTA measurements.

At the top of the experimentally simulated boundary layer, the non-dimensional power spectral densities show an almost perfect overlap with the Eurocode reference spectrum. This consistency across all tested *RH* values indicates that the boundary layer developed naturally and fully over the tunnel fetch, regardless of the roughness height imposed.

4 Conclusions

Experimental investigations were conducted in the downwind test section of the TASL1-M BLWT to determine the distributions of velocity, turbulence intensity, and power spectral density. For this purpose, a 1D wire probe connected to a CTA was used.

Two sets of measurements were performed. The first set featured very high spatial resolution, with 80 measurement points spaced either 1 mm or 5 mm apart, depending on the distance from the wind tunnel floor. The *RH* parameter was set to 0 mm, and the upwind turbulence generators were removed, corresponding to a flow over a flat plate. The post-processed data were fitted to the law of the wall, and a value of 5.3378 resulted for the *B* constant, which showed very good agreement with theoretical predictions.

The second set of measurements used a lower spatial resolution, but with improved time resolution. The acquisition frequency, *f*, was increased to 4 kHz, instead of 1 kHz as it was used in Set 1. Also, the measurement duration *T* was extended from 60 seconds to 120 s. For these tests, the spires and castellated barrier were reinstalled upstream of the boundary layer development zone.

The data from Set 2 was used to determine velocity and turbulence intensity profiles, aligned with the specifications of Eurocode EN 1991-1-4:2005. To achieve this goal, different values of the roughness height (*RH*) were used to experimentally simulate each terrain category.

The results demonstrate good agreement between the measured velocity profiles and the code specifications across various terrain categories.

Turbulence intensity profiles also closely follow the code-defined distributions. However, for terrain categories characterized by higher aerodynamic roughness, a deviation is observed near the tunnel floor, where the predicted values are smaller than those given by the code. This underestimation may be attributed to a low local Reynolds number in the near-wall region.

The power spectral density of the longitudinal velocity component at the top of the boundary layer matches the Eurocode reference spectrum with high accuracy. This consistency was observed across all tested *RH* values.

The findings validate the effectiveness of the variable roughness system in quickly producing realistic boundary layer flows for different exposure factors. Future studies may focus on optimizing *RH* distributions upwind the downwind experimental zone, and increasing flow velocities to refine the simulation of near-ground turbulence intensity further.

References

1. Cermak, J.E.: Wind-tunnel development and trends in applications to civil engineering. J. Wind Eng. Ind. Aerodyn. **91**, 355–370 (2003)
2. Blocken, B., Stathopoulos, T., Carmeliet, J.: CFD simulation of the atmospheric boundary layer: wall function problems. Atmos. Environ. **41**(2), 238–252 (2007)
3. Schlichting, H., Gersten, K.: Boundary-Layer Theory, 9th edn. Springer, Berlin, Heidelberg (2017)
4. European Standard. Eurocode 1: Actions on structures—Part 1–4: General actions—Wind actions EN 1991-1-4:2005, 1–147(2010)
5. Kozmar, H.: Truncated vortex generators for part-depth wind-tunnel simulations of the atmospheric boundary layer flow. J. Wind Eng. Ind. Aerodyn. **99**, 130–136 (2011)
6. Giger, I., Coșoiu, C.I., Georgescu, A.-M.: Experimental investigation of longitudinal velocity distributions in a boundary layer wind tunnel. In: 16th International Conference on Wind Engineering ICWE16, Florence, Italy. https://www.icwe2023.com/. Accessed 10 Nov 2024 (2023)
7. Sánchez-García, R., Gómez-Martínez, R.: Turbulent flow over a rough surface in a wind tunnel. Wind. **5**(2), 13 (2025)
8. Nahian, Z.A., An, L.-S., Fernández-Cabán, P.L., Jung, S.: Artificial neural networks for predicting mean wind profiles over heterogeneous terrains. J. Wind Eng. Ind. Aerodyn. **257**, 105969 (2025)
9. Romanic, D., Kassab, A., Chowdhury, J., Hangan, H., Doddipatla, L.: An analysis of the influence of a generic building on tornadic flow fields using high-frequency PIV and point velocity measurements. J. Fluids Struct. **123**, 104010 (2023)
10. Haselgrove, J., Kelso, R., Emes, M.: Development of a low-cost four-hole pressure probe for the measurement of three-dimensional fluid flows. In: 23rd Australasian Fluid Mechanics Conference, Sydney, Australia (2022)
11. Canepa, F., Burlando, M., Romanic, D., et al.: Downburst-like experimental impinging jet measurements at the WindEEE dome. Sci Data. **9**, 243 (2022)
12. Bode, F., Joldos, T., Sîrbu, G.M., Danca, P., Năstase, I., Coșoiu, C.I.: Impact of realistic boundary conditions on CFD simulations: a case study of vehicle ventilation. Build. Environ. **267**(A), 112264 (2025)

13. Bütefisch, K.A.: Three-component laser doppler anemometry in large wind tunnels. Prog. Aerosp. Sci. **26**(1), 79–113 (1989)
14. Coghe, A.S., Cossali, G.E.: Effect of validation procedure on step noise amplitude in LDV spectra evaluation. Exp. Fluids. **24**, 382–388 (1998)
15. Jedelsky, J., Frantiesk, L., Jicha, M.: Power spectral density of velocity fluctuations estimated from phase Doppler data. EPJ Web Conf. **25**, 01034 (2012)
16. Jørgensen, F.E., How to measure turbulence with hot-wire anemometers—a practical guide, Dantec Dynamics A/S (2005)
17. Erdil, A., Arcaklıoğlu, E.: Hot wire probe calibration using artificial neural network. Trans. Inst. Meas. Control. **31**(2), 153–166 (2009)
18. Vlăduţ, A.C., Popa, S.I., Coşoiu, C.I., Georgescu, A.M., Degeratu, M., Haşegan, L., Anton, A.: A new boundary layer wind tunnel. UPB Sci. Bull. Ser. D Mech. Eng. **79**(2), 159–168 (2017)
19. White, F.M., Majdalani, J.: Viscous Fluid Flow, 4th edn. McGraw-Hill, New York (2022)
20. Hewes, A., Medvescek, J.I., Mydlarski, L., Baliga, B.R.: Drift compensation in thermal anemometry. Meas. Sci. Technol. **31**, 045302 (2020)

MIX
Papier aus verantwortungsvollen Quellen
Paper from responsible sources
FSC® C105338

If you have any concerns about our products,
you can contact us on
ProductSafety@springernature.com

In case Publisher is established outside the EU,
the EU authorized representative is:
Springer Nature Customer Service Center GmbH
Europaplatz 3, 69115 Heidelberg, Germany

Printed by Libri Plureos GmbH
in Hamburg, Germany